Structural Geology

An Introduction to Geometrical Techniques

Structural Geology

An Introduction to Geometrical Techniques

Second Edition

DONAL M. RAGAN
Department of Geology
Arizona State University

John Wiley & Sons.

New York   London   Sydney   Toronto

Library of Congress Cataloging in Publication Data

Ragan, Donal M.
 Structural geology.

 Bibliography: p.
 1. Geology, Structural.
QE601.R23   1973   551.8   73-3335
ISBN 0-471-70481-4

10 9 8 7

Preface

To achieve a high level of understanding in a short time it is necessary to master the essentials of both the traditional approaches and recent advances. In structural geology, the problem of doing this is compounded partly because of the revolutionary character of many of the modern developments and partly because there are few short-cuts to the all-important spatial visualization.

The first steps in the study of geologic structures are largely geometrical. This was true in the historical development of the field, and it is also true in the initial stages of an investigation and in the education of a structural geologist. This concern for geometry includes the basic methods of describing and illustrating the form and orientation of geologic structures, and the solution of various dimensional aspects of structural problems. However, geometry is not the end. The final goal is a full mechanical understanding of the products and processes of rock deformation.

This book attempts to fill a need for a modern introduction to geometrical techniques used in structural geology. The contrast with the more traditional approaches lies in an emphasis on visualization and on those aspects of geometry which yield insights into important new and old problems. Not everything could be included, and it has been necessary to trim some of the material usually covered. Even here, however, I believe that the essence of the older, well-established techniques has been preserved. Where possible, I have tried to bridge the gap between the old and the new by blending the two. I have as well included material not commonly used at the introductory level; in particular, certain fundamental mechanical concepts are applied to the interpretation of simple structures. Not only are these applications important in their own right, but even more they indicate where many of the most challenging problems of structural geology lie. Students who go no further in structural geology should have a working knowledge of the basic geometrical techniques and should be able to follow much of the specialized literature with at least some understanding. At the same time, those who do go on, either in advanced courses or on their own, should have the necessary foundation.

In common with most treatments, the first few chapters apply the methods of orthographic projection to the solution to simple structural problems. For those who feel the need for an introduction or review of descriptive geometry, an outline is given in Appendix A. Use is made in these early chapters, as well as in later ones, of Mackin's powerful method of visualization—the down-structure view of geologic maps.

Fundamental aspects of progressive strain in two dimensions are introduced with the use of card deck models, and the concepts derived from these models are then applied to the analysis of several geologic problems. The difficulties of describing the state of inhomogeneous strain leads to the geometrical description and classification of folds. The properties of ideal parallel and similar folds are explored in some detail, again with the aid of card deck models. The opportunity presented by the shear fold mechanism is used to introduce superposed folding.

The stereographic projection is introduced fairly early, and most of the subsequent problems are solved with it. The use of the stereonet in structural analysis follows naturally, and the method of constructing the various types of diagrams, including contouring, are described.

Faults are described and classified, and problems of displacement are solved by combining orthographic and stereographic methods. The geometrical aspects of stress are developed in much the same terms as the earlier analysis of finite strain in order to emphasize the formal similarity. These concepts, together with the Coulomb criterion of shear failure are then applied to the analysis of the faults.

Finally, methods of presentation and analysis of geologic data, including maps, structure sections, and block diagrams are developed. Throughout, emphasis is given to the geologic map because it is surely the single most important tool in structural geology.

Definitions of a number of important terms are given at the beginning of most of the chapters. Generally, I have followed accepted usage, and for the most part the definitions are taken from the *International Tectonic Dictionary* (Dennis, 1967). However, I have not hesitated to modify these definitions in order to achieve greater clarity. Nor have I attempted to discuss alternative or conflicting usages; these can be found in the Dictionary or in structural geology texts.

A book such as this is by its very nature eclectic, and I hasten to acknowledge my debt to the large number of authors whose material I have adapted. I also thank the various publishers for giving permission to reproduce copyrighted illustrations; these sources are noted at appropriate places throughout the text. I especially thank the Journal of Geological Education for permission to use the material appearing as Chapter 5. If there is a single recurring theme in the book it is the down-structure view of maps, and I owe my gratitude to the late J. H. Mackin for a thorough grounding in the method. A large number of readers and reviewers of the first edition made many important comments, suggestions and corrections; I can not list all their names here but this does not diminish my thanks to them. S. L. Anderson, D. J. Fisher, N. J. Price, J. C. Rosner, and M. F. Sheridan each read one or more of the chapters and their comments lead to noticeable improvements. J. G. Ramsay's influence will be seen on many pages; he also kindly supplied Figure X7.1. I am of course responsible for any remaining shortcomings and inevitable errors. Once again, I should appreciate hearing about them as they are found.

Donal M. Ragan
Tempe, Arizona

Note to the Student

The proof of these geometric techniques is in their application. To this end some exercise material is included with each chapter. However, an important part of structural geology is the study of geologic maps. Such maps cannot be included here, but many excellent ones are readily available. U.S. Geological Survey maps, both old and new series, provide a wide variety of examples from many areas. Some of the state geological surveys publish good maps. Occasionally the Geologic Society of America Bulletin contains maps in full color, and numerous black and white maps appear in this and other journals. Two collections of geologic maps with accompanying sections and explanatory text are commercially available (see *Suppliers*). Several books by British authors contain a number of maps and map problems (see *Bibliography of Geometrical Techniques*, p. 201).

The accompanying exercises, and map problems generally, can usually be solved with a minimum of equipment and materials. The required equipment should include a drafting-type compass, a semicircular protractor, an accurate scale, and a straight edge (T-square or large triangle). Those contemplating the purchase of these items or those using drafting equipment for the first time are advised to read the section in Appendix A on maintaining accuracy in drawings. As with all things a better job can be done quicker with more elaborate and specialized equipment. A drafting machine is an especially desirable addition.

Printed stereographic projections, other graphic aids, and some exercise material are grouped at the back of the book on perforated pages. It is convenient to permanently mount the two 15 cm stereonets. One successful method is to glue the printed nets, being careful not to alter their dimensions, to each side of a 20 cm square of pressed sawdust board (also called particle board) which has been well coated with shellac. The surface of the nets can then be protected with a sheet of self-adhesive clear plastic. A small hole is made exactly at the center to accept a map pin.

One important skill that should be cultivated along with an increased understanding of structural geometry is the ability to produce an effective diagram. The requirements are, in varying degrees, both technical and artistic. For many purposes a technically competent diagram is adequate; the literature is full of examples. With care and a little experience, the necessary attributes of clarity and accuracy can be developed. However, an artistic touch invariably raises the quality of an illustration above the merely adequate. Despite disclaimers, at least some of this skill can be acquired and the student is urged to practice by making quick, three-dimensional sketches of various structural features as they are encountered. The importance of developing this habit cannot be overemphasized. By visualizing the three-dimensional form of a structure and then making that visualization concrete by drawing, the ability to do both is greatly strengthened.

SUPPLIERS

Geologic maps available from
Williams and Heintz Map Corporation
8351 Central Avenue
Washington, D. C. 20027

1. Geologic Map Portfolio No. 1, compiled and edited by L. W. Currier.

2. Geologic Map Portfolio No. 2, compiled and edited by Forbes Robertson.

Stereonets

1. Excellent, but expensive 20 cm Wulff and Schmidt nets are available from E. Leitz, Inc., Rockleigh, New Jersey 07647.

2. 10 cm and 20 cm Wulff and Schmidt nets of very good quality at modest individual and bulk prices are available from the Geological Journal, Department of Geology, The University, Liverpool, England.

3. Very inexpensive 10 cm and 18 cm Wulff nets of good quality are available from the Bookstore, University of Chicago, Chicago, Illinois 60637

Plastic sheets

4. Self-adhesive plastic protecting sheets are commonly available in stationery stores and household departments, including:
1. Cook's Seal-Vu laminating film
2. Stenso Clear Seal
3. Transparent Con-Tact

Contents

1
Attitude of Planes

DEFINITIONS

ATTITUDE

The general term for the orientation of a structural plane or line in space, usually related to geographic coordinates and the horizontal. Both bearing and inclination are components of *attitude*.

BEARING

The horizontal angle between a line and a specified coordinate direction, usually true north or south.

INCLINATION

The general term for the vertical angle, measured downward, between the horizontal and a plane or line.

STRIKE

The bearing of a horizontal line on an inclined plane (Fig. 1.1).

DIP

The inclination of the line of greatest slope of an inclined plane. It is measured perpendicular to strike (Fig. 1.1).

APPARENT DIP

The inclination of a plane measured in a direction *not* perpendicular to the strike (Fig. 1.2).

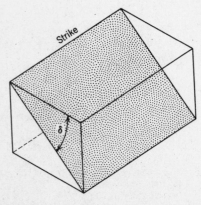

FIGURE 1.1 Angle of true dip δ measured in vertical plane perpendicular to the strike.

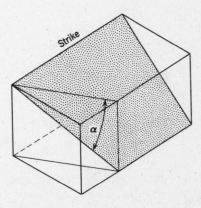

FIGURE 1.2 Angle of apparent dip α measured in vertical plane not perpendicular to the strike.

DIP AND STRIKE

The terms dip and strike apply to any planar structure, and together their values constitute a statement of the attitude of the plane in space. The planar feature most frequently encountered in many areas is the bedding plane; it is also the one dealt with most in beginning structural geology. Other structural planes include cleavage, schistosity, and joints.

Each type of plane has a special map symbol, called a dip and strike symbol, which in general has three parts:

1. *Strike line*, which should be plotted long enough so that the bearing can be accurately measured again from the map.
2. *Dip mark*, at the midpoint of one side of the strike line, indicates the direction of downward inclination of the plane.
3. A number close to the dip mark, and on the same side of the strike line, indicates the dip angle.

The only exceptions are the special cases of horizontal and vertical attitudes. The most common symbols are given in Fig. 1.3, and their use is well established by convention. However, it is sometimes necessary to use these or other symbols in special circumstances, so that their exact meaning should always be made clear in the map legend. The values of dip and strike angles are also often referred to in text. This usage is considerably less standard. Each of the following four examples refers to exactly the same attitude:

1. N 65 W, 25 S: strike is 65° west of north, and dip is 25° in a southerly direction.
2. 295, 25 S: strike is 295° measured clockwise from north, dip is to the south.
3. 25, S 25 W: dip direction has a bearing of 25° west of south, inclination is 25° in this direction.
4. 25, 205: dip direction has a bearing of 205° measured clockwise from north, dip is 25° in this direction.

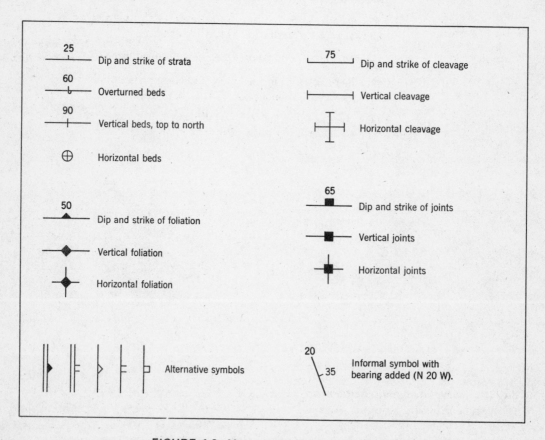

FIGURE 1.3 Map symbols for structural planes.

The first two forms are the most common with the difference usually depending on whether the compass used to make the measurement is divided into quadrants or the full 360°, and on personal preference. The advantage of the quandrant method of presentation is that it is much easier to quickly grasp a mental picture of the attitude, and will therefore be used in this text. The third, and closely related fourth form will be used for the bearing and inclination of lines rather than planes, although sometimes those lines will lie in planes, such as dip lines, in which case both forms may apply. Again the quadrant form will be used here, although the fourth method gives the attitude of the plane unambiguously without need of letters for geographical directions, and is therefore particularly useful in computer treatment of attitude data.

There are a number of ways to determine the attitude of structural planes. All are based on field measurements of one kind or another. The most direct method is to hold the compass directly against the bedding or other plane surface at the outcrop. One edge of the open compass case is placed against the surface and the compass rotated until it is level (Fig. 1.4a). The bearing in this position is the strike. Dip is determined by placing one side of the compass box and lid directly on the exposed plane at right angle to the previously measured strike direction. The clinometer bubble is leveled and the dip angle read (Fig. 1.4b). While simple to use, this method is subject to rather large errors because of the short compass base and the imperfection of naturally occurring plane surfaces. Other more accurate direct measurement techniques can be found in any field geology text (e.g. Compton, 1962).

Indirect methods are also available, and are the subject of this and a later chapter. All the methods dealt with here are concerned with determining the relationship between the components of the attitude—the apparent dip, true dip and strike.

GRAPHIC METHODS

From the point of view of visualizing, and therefore thoroughly understanding the spatial relations of the various angular components, wholly graphic constructions are the most important. Once the ability to visualize is attained, other quicker, more efficient methods can be applied.

The simplest type of problem involves determining one component from the other two.

PROBLEM

Given the attitude of a plane (N 90 E, 30 N), find the apparent dip in the vertical section trending N 45 W.

APPROACH (Fig. 1.5)

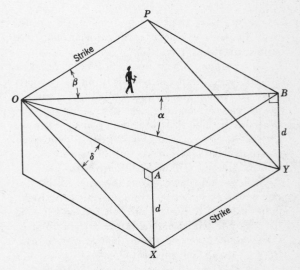

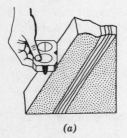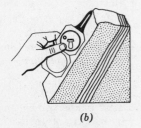

(a)   (b)

FIGURE 1.4 Measurement of strike (a) and dip (b) with a Brunton compass. (From Compton, 1962, Manual of Field Geology, Wiley, used by permission.)

FIGURE 1.5 Block diagram showing the angles involved in problems of true dip and apparent dip.

From one line of strike on the plane (OP), and the true dip, a simple triangle (OAX) gives the location of a second line of strike (XY) with known elevation relative to the first. Such lines on a plane are called structure contours, and the vertical distance between two adjacent contours is the contour interval (here d). Moving on a horizontal map surface obliquely from one structure contour to the other (in the example from O toward B) the depth to the plane steadily increases; at point B on the second contour it is exactly d. A second triangle can therefore be constructed which gives the inclination of the plane in this direction, that is, the apparent dip in a direction with the specified bearing.

CONSTRUCTION (Fig. 1.6)

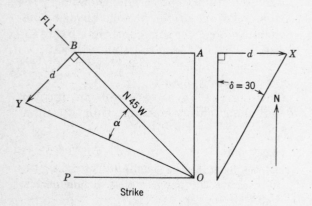

FIGURE 1.6 Graphical determination of apparent dip from known true dip and strike.

1. In map view draw a line of strike (OP), and a line of some convenient length perpendicular to it representing the dip direction (OA).
2. Draw a second strike line (AB) through point A.
3. To find the difference in elevation (d) between the two contours, construct a right triangle using OA as one side and the angle of true dip (δ). The other side is then the required d, giving point X on the second line of strike. Note that this triangle can be constructed independently of the map view.
4. Add the N 45 W line (OB) to the map.
5. The second triangle giving the apparent dip (a) can now be constructed. In doing this it is always more efficient to use line OB as a folding line (FL 1). The side with length d is drawn from point B, giving point Y on the second line of strike. Angle BOY is the required apparent dip.

ANSWER

For a plane with attitude N 90 E, 30 N, the apparent dip in the N 45 W direction is 22°.

A similar construction is involved in determining the attitude of a plane from two apparent dip measurements.

PROBLEM

Given two apparent dips (10, N 72 W, and 25, N 35 E) obtain the true dip and strike of the plane.

APPROACH (Fig. 1.7)

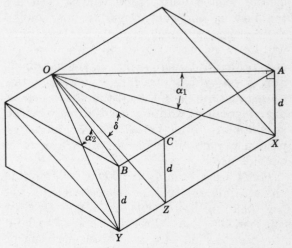

FIGURE 1.7 Block diagram showing the angles involved in the problem of determining dip and strike from two apparent dips.

The apparent dip measurements represent the attitudes of two lines which intersect on the plane of interest. Plotting the bearings of these lines established this point in map view. Since three points determine a plane, two more must be found. A second point is found from a triangle plotted from the information of one apparent dip measurement (OAX). Another point, associated with the second apparent dip could be found in like manner. However, it is advantageous to locate this third point at the same relative elevation as the second. The line joining these two points of equal elevation is then a structure contour, or line of strike. The dip is measured perpendicular to this strike line.

CONSTRUCTION (Fig. 1.8)

1. Plot the two apparent dip bearings in map view, intersecting at point O.
2. Using one of these lines as a folding line (FL 1) and the associated apparent dip construct triangle OAX. The lengths of OA and depth d are determined by convenience only, but should be of reasonable length to insure accuracy.

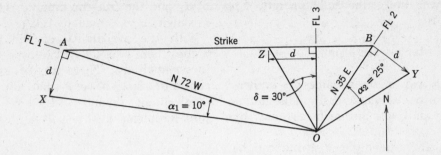

FIGURE 1.8 Graphical construction of dip and strike from two known apparent dips.

3. Using the second of these apparent dip bearings as a folding line (FL 2) and its apparent dip, construct another triangle *OBY*. This time, however, the depth *d* muxt be *identical* with that used in step 2.

4. Two points (*X* and *Y*) of equal (but unknown) elevation are now located; their positions in map view are *A* and *B*. To fix the strike direction join points *A* and *B*.

5. A line from point *O* perpendicular to this strike line (*AB*) determines the dip direction. With this line as folding line (FL 3) scale off the same depth *d*, thus establishing point *Z* on the line of strike.

6. A line joining points *O* and *Z* determines the slope of the plane in the direction perpendicular to the strike, that is, the true dip.

ANSWER

The attitude of the plane represented by the two apparent dips is N 90 W, 30 N.

This same type of problem may be solved by a short-cut method which combines the principal features of the full graphic method with trigonometric data.

CONSTRUCTION (Fig. 1.9)

1. As before plot two rays in map view from a single point representing the bearings of the apparent dip measurements.

2. With the use of any convenient scale plot the distances equal to the value of the cotangent of each apparent dip along the respective rays, starting at the point of intersection.

3. A line joining the two points so determined is the line of strike.

4. The perpendicular distance from the intersection to this strike line is equal to the cotangent of the true dip angle.

TRIGONOMETRIC METHODS

The value of the methods of descriptive geometry lies in the practice in visualization that comes with manipulating the elements of a problem in three dimensions. The importance of developing this ability to visualize can not be emphasized enough. It will be obvious, however, that these orthographic constructions have practical limitations. For this reason, alternative methods have been devised to quickly and easily solve the various attitude problems.

Since the graphical solutions of dip and strike problems involve a series of right triangles, it is clear that solutions can also be obtained trigonometrically. The relationship linking true dip δ, apparent dip a, and the

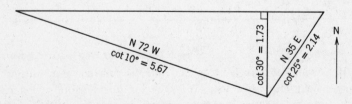

FIGURE 1.9 Graphic-trigonometric method for determining dip and strike from two apparent dips.

angle between the strike and apparent dip direction β, is

$$\tan a = \tan \delta \sin \beta \qquad (1.1)$$

Thus, if only two of the variables are known, the third can be calculated easily. This can be done simply and with sufficient accuracy by slide rule.

Similarly, the true dip and strike can be found from two apparent dips from the following equations (after Earle, 1934, p. 2; Hughes, 1960):

$$\tan \phi = \left\{ |(\tan a_2 / \tan a_1) - \cos \theta| \right\} / \sin \theta$$
$$\tan \delta = \tan a_1 / \cos \phi$$

where a_1 and a_2 are the apparent dip angles, θ the angle between the two apparent dip bearings, and ϕ the angle between the bearing

of a_1 and the true dip direction (see Fig. 1.7; θ = angle AOB, ϕ = angle AOC).

With the availability of still other techniques there is usually little need to resort to these equations. There are two notable exceptions: first, if very small dip angles are involved, and second, where large numbers of these problems must be solved routinely.

OTHER TECHNIQUES

A variety of graphical and mechanical aids based largely on the trigonometric relationships have been developed. An alignment diagram (Fig. 1.10) is especially easy to use. A simple rule based on cotangent values (White, 1946) and a nomogram (Leney, 1963) may also be used to solve dip problems, as well as a variety of other problems involving angles. Or

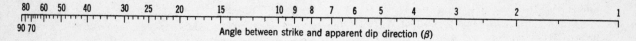

Angle between strike and apparent dip direction (β)

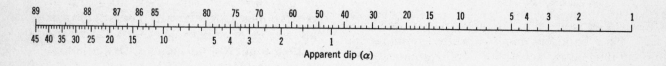

Apparent dip (α)

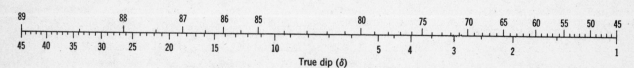

True dip (δ)

FIGURE 1.10 Alignment diagram for determining apparent dip. (From Nevin, 1949, Principles of Structural Geology, Wiley, used by permission.) Using a straight edge join the points representing the angle between the strike and the apparent dip direction on the first column and the true dip on the third column. The apparent dip is read off the middle column.

a special circular slide rule (Satin, 1960) and a protractor (Ten Haaf, 1967) may be used.

However, the most useful technique of all is based on a completely different type of projection and a specially constructed plotting net. Not only can these, as well as a wide variety of other problems involving even more complex angular relationships be treated, but completely satisfactory numerical solutions can be obtained very quickly, and the entire plotting process checked by visualization, thus making it an exceedingly powerful technique. The details of this method are described at length in Chapter 11.

EXERCISES

A word of advice before you start your first graphical constructions—you will be repaid many times over if you develop the habit of sketching out the main elements of the problem before you attempt an accurate scale drawing. The first advantage is that the sketch will allow you to place the final drawing on the sheet of paper more efficiently. It is a common, but annoying experience to start the drawing on the middle of the sheet only to find that the construction takes you off the page. Even more important, you will find that it is relatively easy, even after considerable experience, to take a wrong turn if you start the full construction first. In concentrating on the accurate plotting, you may measure an angle in the wrong direction from north, or plot the dip on the wrong side of the strike line, or measure a distance along the wrong line. The quick preliminary sketch will encourage you to see the forest, not just the trees.

If you find the use and meaning of folding lines troublesome, it often helps to literally bend the paper around the edge of the table top along the required line of your sketch. You can then examine the map view and vertical section simultaneously. Having seen these views, it is usually an easy matter to decide just how the final drawing should proceed.

1. Become familiar with the use of a Brunton compass. Especially if the dial is divided into quadrants, make sure that you understand how to measure bearings. Because the axis of the compass case is aligned parallel to the structure to be measured, not the needle, the E and W are reversed from their normal positions, and this may lead to some confusion for the beginner. Note also that the dial can be adjusted so that the compass reads *true* rather than magnetic north.

2. For the following attitude data, determine the unknown component three different ways: (i) graphically, (ii) trigonometrically, and (iii) with the alignment diagram. For one problem, solve graphically using two different scales—the second twice the first. In the field, angles can be measured with a compass and clinometer to an accuracy of no better than one degree, if that. What then can you say about the relationship between the scale of your two constructions and the justifiable accuracy?

   (a) If the attitude of a plane is N 75 E, 22 N, what is the apparent dip in the direction N 50 E?

   (b) An apparent dip is 33, N 47 E; the strike is E-W. What is the true dip?

   (c) The true dip is 40° due north. In what direction will an apparent dip of 30° be found?

3. Graphically determine the true dip and strike from each pair of apparent dip measurements. For one pair also calculate the attitude trigonometrically.

   (a) 20, N 80 W, and 40, N 10 E

   (b) 30, N 60 E, and 50, S 45 E

   (c) 6, N 78 W, and 25, N 36 W

4. A smooth, vertical cliff exposes the traces of uniformly dipping strata. There is no hint of the third dimension. What is the appearance of these inclined traces as viewed from different positions on the ground? What can be said about the true attitude of the beds with no further data?

5. Probably the most important application of the technique for determining the angle of apparent dip from true dip and strike is in the construction of vertical structure sections. Fig. X1.1 shows the edge of a proposed trench 10 m deep. The top of a distinctive sandstone unit outcrops at point A, and has an attitude of N 60 E, 40 N. The area is also cut by a large number of joints (N 20 W, 65 W). Draw a diagram to scale showing the vertical north wall of the trench as it will appear when excavated. Include point A on your section, and show the joints diagrammatically.

6. Three points A, B, and C on an inclined stratum have elevations of 150 m, 75 m, and 100 m, respectively. The distance from A to B is 1100 m in a direction of N 10 W, and from A to C is 1560 m in a direction N 40 E. What is the dip and strike of the stratum?

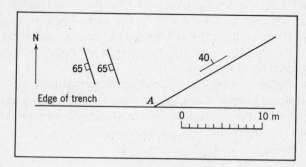

FIGURE X1.1

2
Thickness and Depth

DEFINITIONS

THICKNESS

The perpendicular distance between the two parallel planes bounding a tabular body of rock (Fig. 2.1). Stratigraphic thickness is a special case.

DEPTH

The vertical distance from a specified level (usually the earth's surface) downward to a point, line or plane (Fig. 2.1).

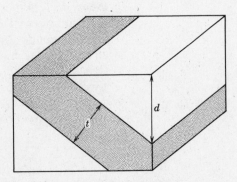

FIGURE 2.1 Block diagram showing thickness t and depth d.

THICKNESS

The thickness of a layer may be determined in a number of different ways (for a full discussion, see Kottlowski, 1965). In favorable situations it may be possible to obtain thickness by direct measurement; otherwise it must be determined indirectly.

9

Direct Measurement. Several examples will illustrate how thickness may be measured directly. In the simplest case, of a horizontal layer exposed on a vertical cliff face, the thickness may be obtained by hanging a tape over the edge of the cliff (Fig. 2.2). Similarly, if the elevation of the top and bottom of the layer can be determined accurately, regardless of the slope angle, the thickness is simply the difference in elevation. Another special case is the exposure of a vertical layer on a horizontal surface; a measuring tape extended perpendicular to the strike gives the thickness (Fig. 2.3).

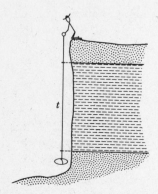

FIGURE 2.2 Direct measurement of the thickness of a horizontal layer.

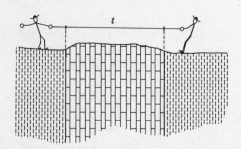

FIGURE 2.3 Direct measurement of the thickness of a vertical layer.

More generally, thickness may be directly measured regardless of the dip-slope relationship with a Jacob's staff (a light pole with gradations and hand level with clinometer or a Brunton compass attached at the top). The clinometer is set to read the dip angle and the staff is inclined in the direction of dip until the bubble is centered. A point on the ground

is then sighted in. The thickness of the bed or portion of a bed between the base of the staff and the sighted points is equal to the height of the staff. This method can be used for thin beds (thickness less than the staff height), or, by occupying successive points, for units of any thickness (Fig. 2.4). The staff must be inclined exactly in the dip direction and perpendicular to the strike, otherwise an incorrect thickness is obtained. The principle common to each of these direct approaches is that a line of sight can be obtained that is parallel to the line of dip, the layer appears in edge view and the thickness can be obtained by measuring across in this view plane perpendicular to the parallel bounding surfaces.

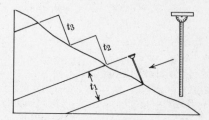

FIGURE 2.4 Direct measurement of an inclined layer with a Jacob's staff.

Indirect Measurements. When direct measurement of thickness is not possible, there are several alternatives. Which one of these is used depends on the field situation, the equipment at hand, the complexity of the structure, and finally, on personal preference. Given a choice, it is, of course, always desirable to use the most nearly direct measurements possible.

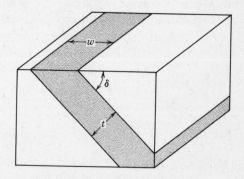

FIGURE 2.5 Block diagram showing outcrop width w.

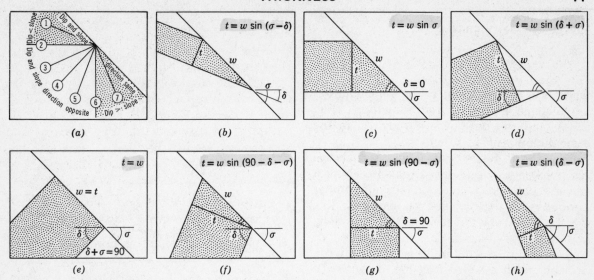

FIGURE 2.6 Determination of thickness from outcrop width measured on a slope.

The simplest of the indirect approaches is to measure the width of the exposed layer perpendicular to the strike on a horizontal plane surface. From this outcrop width w and the angle of dip δ, the thickness t can be determined by constructing a scaled triangle, or by using the following equation (see Fig. 2.5):

$$t = w \sin \delta \qquad (2.1)$$

Essentially the same method may be used when the outcrop width is measured on sloping ground. In this case, the thickness is a function of both the dip angle δ and slope angle σ, with the results that the determination is more involved than in the simpler case of a horizontal surface. Again, the thickness may be found graphically with a scaled drawing, or it may be computed trigonometrically. Because there are three separated measurements involved, together with the possible variations in slope and dip directions,

seven subcases arise. They are shown diagrammatically in Fig. 2.6a, and the individual cases together with the appropriate equations are shown in Fig. 2.6b-h. The geometry of each subcase should be examined in order to understand the variations involved, but once the possible relationships between dip and slope angles are seen, all can be solved easily using equation (2.1) by substituting the appropriate angle in place of dip, without attempting to memorize each separate case.

A second approach, after Secrist (1941), involves the measurement of two different components: the horizontal h and vertical v distances between two points along a line perpendicular to the strike. Because slope angle and slope distance are not used, this is a convenient method when irregular slopes are involved. It can also be used to determine the thickness of units from geologic maps. The three principal cases are illustrated in Fig. 2.7.

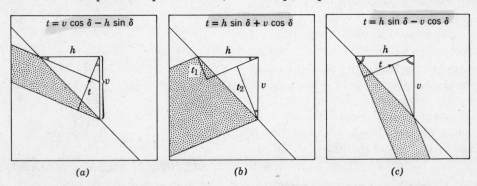

FIGURE 2.7 Thickness from vertical v and horizontal h distances instead of slope measurements.

The general equation is:

$$t = h \, |\sin \delta \pm v \cos \delta| \qquad (2.2)$$

where the sum is taken if the slope and dip are inclined in the opposite directions (Fig. 2.7b), and the difference if they are in the same directions (Fig. 2.7a,c).

Because of obstructions, it may not always be possible to make measurements along a traverse perpendicular to the strike. Another correction is then needed, as shown in Fig. 2.8. In effect, the traverse length l is too long and must be reduced to the equivalent of outcrop width. As before, this correction can be made with a scaled drawing, or with the equation:

$$w = l \sin \beta \qquad (2.3)$$

where β is the angle the traverse direction makes with the strike. Or the thickness may be determined with the equation:

$$t = l \sin \beta \sin \delta \qquad (2.4)$$

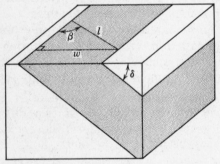

FIGURE 2.8 Block diagram showing a measured traverse oblique to the strike.

In the more general case of an oblique traverse made on a slope, a similar, though more complicated, relation exists. The complication arises from the usual restriction of making only vertical or horizontal angular measurements in the field from which it is not possible to obtain the outcrop width, or its equivalent, as in equations (2.3) and (2.4). Fig. 2.9 illustrates the problem. The appropriate trigonometric equation in this situation is (after Mandelbaum and Sanford, 1952, p. 767):

$$t = l \left\{ | \sin \delta \cos \sigma \sin \beta \pm \sin \sigma \cos \delta | \right\} \ (2.5)$$

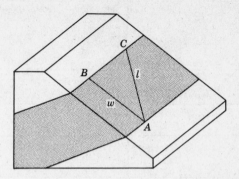

FIGURE 2.9 Block diagram showing an oblique traverse made on a slope. None of the angles of triangle *ABC* are directly measurable in the field.

The sum is taken when the slope and dip directions are opposite, and the difference when they are in the same direction.

DEPTH

Once the three-dimensional relationships involved in the determination of thickness can be visualized, and the methods used in the calculation understood, the solution of depth problems will present no difficulty. As with thickness, the simplest case is the depth d to an inclined plane exposed on a horizontal topographic surface (Fig. 2.10). Given the dip angle and the horizontal distance m measured perpendicular to the strike from the outcrop to the point of interest, depth may be determined by constructing the appropriate right triangle, or by using the formula:

$$d = m \tan \delta \qquad (2.6)$$

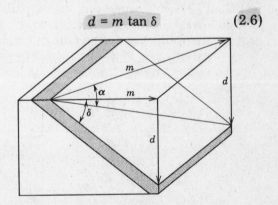

FIGURE 2.10 Block diagram showing depth d from a horizontal surface.

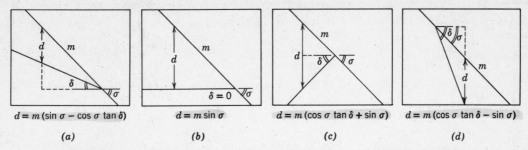

$$d = m (\sin \sigma - \cos \sigma \tan \delta) \qquad d = m \sin \sigma \qquad d = m (\cos \sigma \tan \delta + \sin \sigma) \qquad d = m (\cos \sigma \tan \delta - \sin \sigma)$$

(a)   (b)   (c)   (d)

FIGURE 2.11 Determination of depth from distance *m* measured on a slope.

If the distance m' is not measured perpendicular to the strike, the apparent dip is used.

The case of depth to an inclined plane from sloping ground is somewhat more involved. Several subcases exist, depending on the slope and dip directions, and each is detailed in Fig. 2.11. Generally, with slope distance *m* measured perpendicular to the strike, the equation is:

$$d = m \left\{ |\sin \sigma \pm \cos \sigma \tan \delta| \right\} \qquad (2.7)$$

where the sum is taken if the dip and slope are inclined in the opposite directions, and the difference if they are in the same directions. If the slope distance is oblique to the strike, apparent dip in the traverse direction must be used.

With map data, or the equivalent, depth may be quite easily solved with equation (2.6), or with a simple alignment diagram (Fig. 2.12). Only the horizontal distance from the outcrop point to the point of interest and the apparent or true dip are needed. Depth relative to the elevation of the outcrop point is then calculated, or obtained directly from the diagram. If the point where depth is required is above or below this point, the difference in elevation is either added or subtracted to give the acutal depth.

EXERCISES

1. The attitude of a sandstone unit is N 65 E, 35 S. A horizontal traverse with a bearing of S 10 E, from the bottom to the top measured 125 m. What is the thickness?

2. A limestone formation is exposed on an east facing slope. Its attitude is N 15 W, 26 W. The traverse length from the bottom made with bearing N 90 W is 653 m, and the slope angle was +15° (upward). What is the thickness of the limestone? What is the depth to the lower boundary from the upper end of the traverse line?

3. The following data is from a geologic map. On a line with bearing S 85 W, perpendicular to the strike of a sill (N 5 W, 38 W), two points are located. The eastern one, at the bottom of the layer, has an elevation of 900 m; the one at the top of the unit has an elevation of 1025 m. What is the thickness of the sill?

4. Determine the thickness of the stratigraphic interval between the lower contact (Station 1) to the upper (Station 5) from the following traverse data.

Station	Attitude	Bearing	Slope distance	Inclination
1–2	N 5 E, 38 E	N 85 E	21.5 m	+12° (upward)
2–3	N 10 E, 42 E	N 60 E	48.3	+ 6
3–4	N 18 E, 48 E	N 80 E	36.9	–10° (downward)
4–5	N 15 E, 45 E	S 65 E	13.0	+18

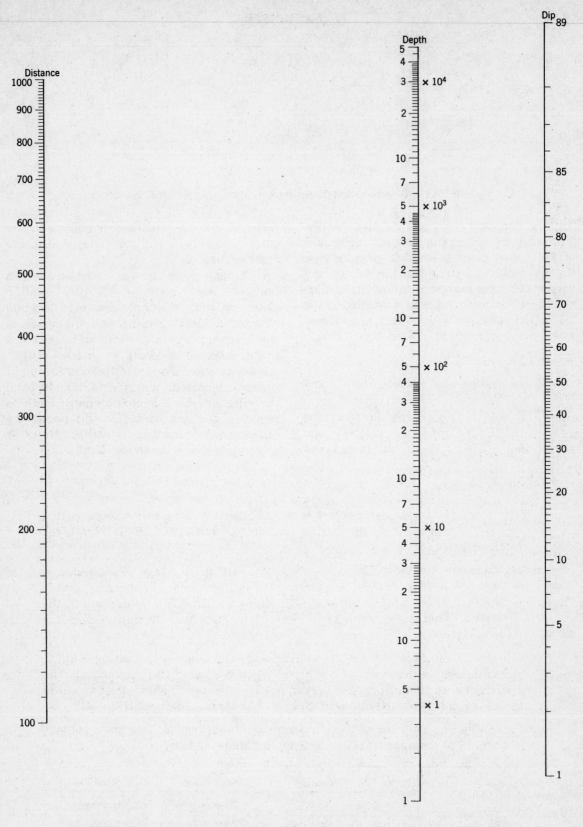

FIGURE 2.12 Alignment diagram for depth problems. (After Palmer, 1918.) With a straight edge join points representing distance and dip; depth appears on the middle column.

3
Planes and Topography

HORIZONTAL SURFACES

To illustrate the methods of determining thickness (Chapter 2), the simplest examples assumed that the earth's surface was a horizontal, geometrically perfect plane. The intersection of inclined planes and layers with this surface results in an *outcrop pattern*, and the trace of this pattern represented in map view is a very simple geologic map. In this case the actual outcrop patterns are limited to straight lines parallel to the strike. The width of the outcrop bands depends on two factors: (1) the actual thickness of the layer, and (2) the angle of dip. The effect of each of these factors is shown in Fig. 3.1. In essence, these relationships also apply to less-than-perfect, real horizontal topographic surfaces.

In the special case of vertical layers, the outcrop width in map view is equal to the thickness of the layer. This unique relationship results from the fact that the map shows the layers in edge view—that is, the line of sight in viewing the map coincides with a line which is parallel to the bounding planes of the layers. In estimating thicknesses, one instinctively seeks just such a line along which to view tabular objects, such as books.

Of the infinite number of such lines, one is always readily identifiable on geologic maps; it is the line of dip. For layers inclined less than 90°, an auxiliary view perpendicular to

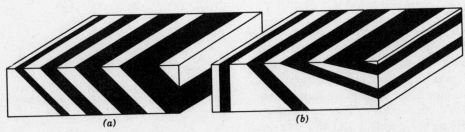

(a)   *(b)*

FIGURE 3.1 Relationship between outcrop width and thickness and dip. (*a*) With dip constant, outcrop width varies with thickness. (*b*) With thickness constant, outcrop width varies with dip.

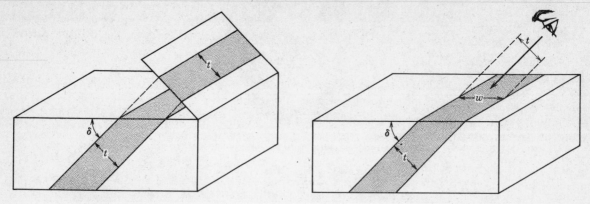

FIGURE 3.2 An auxiliary view showing true thickness. (*a*) Constructed perpendicular to the dip direction. (*b*) Obtained by a down-dip view of the map.

this line could easily be constructed that would also show the layers in edge view, and therefore with true thickness (Fig. 3.2*a*). However, it is unnecessary to make this construction because the same information can be obtained directly from the map by rotating the map and adopting an inclined view of the map surface parallel to the line of dip. In this position, the outcrop width, which is greater than thickness on the map, is foreshortened so that it appears in true thickness (Fig. 3.2*b*). In adopting this oblique view of the map, it helps to reduce depth perception by closing one eye. Clearly, this method is limited to examples involving significant inclination, as it is not physically possible to view horizontal strata in edge view by map rotation.

The same principle is used in reverse for traffic signs painted on streets. By purposely distorting the letters as viewed vertically (map view) the foreshortening that accompanies the driver's oblique view of the road compensates for the distortion and the warnings appear in normal proportions and are perfectly readable (see Fig. 3.3).

In effect, twisting such simple geologic maps so that inclined strata are viewed down-dip rotates the beds back to their original horizontal position. Contacts on maps then cease to be just lines separating stratigraphic units on the earth's surface. They can actually come to life as the depositional and erosional surfaces they are. A map viewed down-dip is a kind of cross section, such as might be seen in the walls of the Grand Canyon, involving an

important additional dimension—sequence of deposition in time. Unconformities become buried landscapes, and this facilitates comparisons with present landscapes, and their erosional processes. This in turn may lead to critical evaluation of relationships and to a more thorough search for new evidence in the field. Certainly the possibility of overturning of the beds should be kept in mind, especially in areas of more complicated structure. In such a case, the down-dip view yields a picture of the strata in a completely upside down position; but such a view may actually help the interpretation of such overturning if other obvious evidence is lacking.

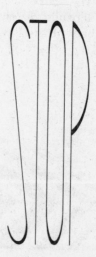

FIGURE 3.3 View with line of sight inclined 20°to the plane of the paper.

RULE OF V'S

In areas of sloping terrain, additional factors are involved in determining the character of the outcrop pattern, including slope angle and direction with respect to the attitude, and on variations in slope angle and direction. In other words, in addition to thickness and angle of dip, map pattern also depends on the details of topography. The relationships between dip and topography have been formalized into a series of rules, called the *Rule of V's*, by which the direction of dip of structural planes can be estimated directly from outcrop pattern. Wherever the trace of a plan crosses a valley the resulting outcrop pattern is characteristic of the attitude, especially dip. There are several distinct type of patterns (Fig. 3.4).

1. *Horizontal planes*: Topographic contour lines can be considered to be the surface traces of imaginary horizontal planes. The traces of real horizontal planes therefore exactly follow the topographic contours. Such patterns are completely controlled by topography; the outcrop trace faithfully reflects the contour lines in every detail. Therefore, the outcrop pattern V's upstream, just as the contour lines do (Fig. 3.4a).

2. *Planes inclined upstream*: As the attitude departs from horizontal, with the dip direction upstream, the pattern made by the horizontal plane is progressively modified into a blunter V, still pointing upstream (Fig. 3.4b). With steepening dip, the outcrop trace is an increasingly subdued reflection of topographic detail.

3. *Vertical planes*: In this special case of a 90° dip, the outcrop trace is straight and parallel to the strike, regardless of the topography. There is no V at all, and thus no control by the topography on the map pattern (Fig. 3.4c).

4. *Planes inclined downstream*: There are three subcases, depending on the relationship between the dip angle and the valley gradient.

   a. With dip greater than valley gradient, the pattern V's downstream (Fig. 3.4d).

   b. If dip and gradient are exactly equal, the outcrop trace does not cross the valley and there is no V (Fig. 3.4e). However, stream gradients steepen headward, continuous planar structure must therefore cross somewhere upstream.

   c. If the dip is less than the valley gradient, but still in a downstream direction, the pattern will V upstream (Fig. 3.4f).

As the valley gradient and structural dip determine whether the pattern will V up- or downstream, the boundary case occurs when the two are parallel (Fig. 3.4e).

These examples illustrate clearly the general principle that map patterns depend on the relation between slope and dip and not directly on the actual slope and dip angles, which are measured with respect to horizontal.

Another and opposite set of rules could be formulated for traces crossing ridges. However, such ridges tend to have greater variety of shapes (sharp, broadly rounded, flat, etc.), depending on the area's geomorphic history, and the patterns are correspondingly variable. Stream valleys on the whole tend to produce more uniform patterns. In the stated rules the strike was assumed to cross the valley at approximately 90°; the resulting V-patterns are more or less symmetrical. With other strike directions, asymmetrical V's result, with the limiting case occurring when the valley axis and strike are parallel.

A simple, easily remembered statement that summarizes all these relationships can be given: the outcrop V points in the direction in which the formation underlies the stream (Screven, 1963). Better yet, however, is to visualize the geometrical relationship between topography and planar structures in three dimensions. In an area of topographic relief the outcrop pattern of dipping beds viewed obliquely from an airplane down the line of dip appears in edge view (the block diagram of Fig. 3.4b is very nearly in this position). The traces of the inclined layers are straight and thickness appears directly. From this position the topographic irregularities have no effect on the pattern, and in this sense it is the same as the case of the vertical dip in map view. In other words, the pattern is complex from every other view point, but simple when viewed down-dip. The relationship would hold, of course, for a scaled topographic relief model with outcrop pattern included. By learning to perceive the surface depicted by contours on a map as a relief model, the mind's eye can accommodate the influence of the topography of the outcrop pattern. This is not always easy to learn; practice is the key. Once the ability is attained, however, it is a powerful aid in map interpretation, for even in areas of considerable and varied relief, and therefore irregular map patterns, the structure can be viewed down-dip directly on the map with a great simplification.

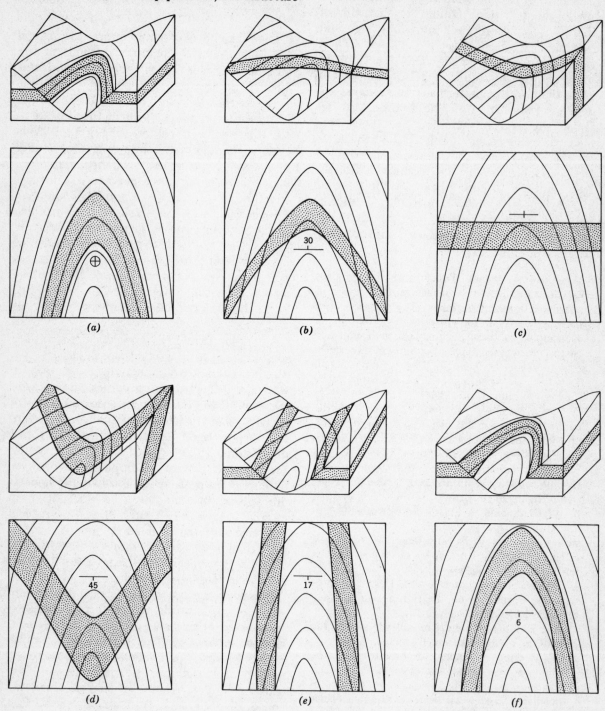

FIGURE 3.4 Outcrop patterns illustrating the rule of V's: (*a*) horizontal layer, (*b*) layer dipping upstream, (*c*) vertical layer, (*d*) layer dipping downstream, (*e*) layer and valley with equal inclinations, (*f*) layer dipping downstream at an angle at less than valley gradient.

DIP AND STRIKE
FROM MAPS

So far the examples have treated the attitude of inclined planes in qualitative terms only. However, the actual values of dip and strike can be found from an outcrop pattern, or its equivalent. All that is required is three points of known position. In the simplest case, the strike is determined by joining two points of equal elevation located on a structural plane. The dip can then be found by measuring the perpendicular map distance and elevation difference from the line of strike to a third point on the same plane, using the following equation:

$$\tan \delta = v/h \qquad (3.1)$$

where v is the vertical distance, or difference in elevation, and h is the horizontal or map distance. Alternatively, a triangle involving the dip angle may be constructed from this same information. The construction involves establishing a folding line perpendicular to the strike and locating the relative elevations of both the strike line and a third point in a vertical plane using the map scale. The dip line can then be drawn, and its angle measured (Fig. 3.5). In either approach, chosing widely spaced points improves accuracy.

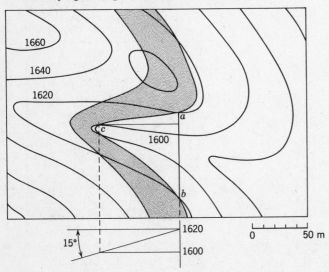

FIGURE 3.5 Dip and strike from outcrop pattern.

THREE-POINT PROBLEM

A more general form of this problem involves determining the dip and strike from three points each with different elevations. Essentially, this is the same problem of determining true dip and strike from two apparent dips. The graphical approach is similar to that of Fig. 1.7, except that it is necessary to construct the apparent dips from given elevations and map distances. Alternatively, the attitude could be determined with the use of a slightly modified version of equation (1.2).

PROBLEM
From the map location and elevations of three points known to be located on a single structural plane, determine the attitude of that plane.

APPROACH
Two points of equal elevations define the strike. In this situation one of these is clearly the point of intermediate elevation. The other point is found between the highest and lowest by dividing the map distance between them into parts proportional to the ratio of the intermediate to one other given point. Calculation of dip follows from the perpendicular distance and elevation difference between the strike line and the lowest point.

CONSTRUCTION
1. If required, plot the three points in map view. Then join them with lines to form a triangle and measure the distance between each point.
2. A point D exists between A and C, the highest and lowest, which has an elevation equal to B, the intermediate. The map distance AD can be determined using a proportional relationship based on two similar triangles:

$$AD/AC = \text{elevation } (D\text{–}A) \text{ / elevation } (C\text{–}A) \qquad (3.2)$$

Or this distance can be determined graphically by constructing these triangles (Fig. 3.6).
3. Line BD is horizontal, and by definition a line of strike.
4. The perpendicular line AE is the dip direction. The elevation of both A and E is known, as is the map distance between them. The dip angle can be calculated using (3.1), or the triangle giving the same information may be constructed (as in Fig. 3.5).

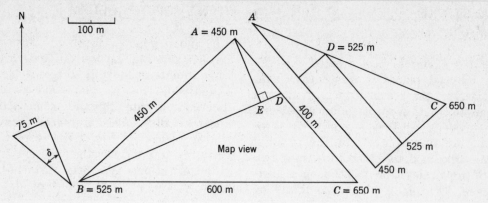

FIGURE 3.6 Three-point problem.

PREDICTION OF OUTCROP PATTERNS

The reverse operation of predicting the outcrop pattern of a structural plane from the attitude at one known point is also possible. If contours are drawn on two intersecting surfaces using a common datum and equal contour interval, contour lines of equal elevation will also intersect. Thus no matter how complicated one or both the surfaces are, their line of intersection can be defined by the points of intersection of like contour lines.

METHOD

For predicting outcrop patterns a map with topographic contours is a necessary prerequisite. It is then necessary to draw these same contours on the structural plane. To do this, the elevations of the contour lines on the dipping plane must be found.

CONSTRUCTION (Fig. 3.7)

1. Draw a line through the single known outcrop point parallel to the strike. This is the first structure contour. The map spacing s of the other parallel contours can be found from $s = i/\tan \delta$, where i is the map contour interval, or as here, graphically.

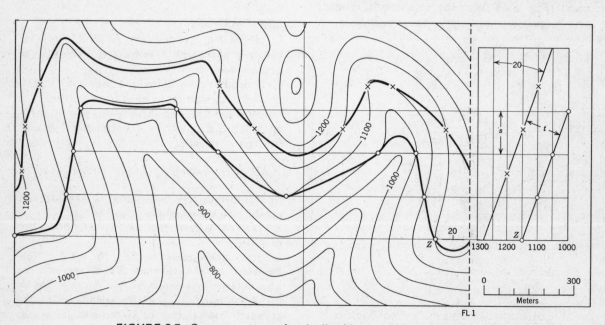

FIGURE 3.7 Outcrop pattern of an inclined layer with base exposed at Z.

2. Using FL 1, perpendicular to this first line of strike, establish a vertical section. In this section:

a. Conveniently locate the known outcrop point Z on the section along the projection line of strike.

b. With the elevation of this point (1150 m) draw a series of horizontal elevation lines equal to the contour interval and at the same scale as the map.

c. Also through point Z draw the trace of the plane inclined at dip angle at 20°.

3. The points where this dip line crosses the elevation lines fix the position of the structure contours on the plane. These are then projected back to the map.

4. Each intersection of a structure contour with its matching topographic contour represents an outcrop point on the line of intersection, and these should be marked distinctly.

5. Complete the outcrop pattern by joining successive outcrop points. This trace must cross or touch the contours at and only at these established points. If the contour spacing is wide, the outcrop trace can often be simply sketched across the gap, keeping in mind the proper relationship between dip and slope. In other cases it may be necessary to use intermediate topographic and structure contours for additional control.

6. If the pattern of a layer is to be predicted, the thickness t is constructed on the vertical section thus establishing the upper bounding plane. The pattern of this second plane is drawn in exactly the same way.

EXERCISES

1. Using Fig. X3.1, determine the dip and strike of the mapped unit. With these results, view the map in a down-dip direction, and in combination with visualization of the topography, try to see the unit as a planar layer in edge view. The topography of Fig. X3.2 is identical. With your visualization of Fig. X3.1 as reference, try to look down the dip of this layer, and estimate its attitude and thickness. Check your results.

2. With the following information, and the topographic map of Fig. X3.3 (see p. X-1 after Index) construct a geologic map. The base of a 100 m thick Triassic sandstone unit is exposed at point A; the attitude is N 70 W, 25 S. Point B is located on the west boundary of a vertical diabase dike of Jurassic age 50 m thick; its trend is N 20 E. At point C, the base of a horizontal Cretaceous sequence is exposed, and at point D the base of a conformable sequence of Tertiary rocks is present.

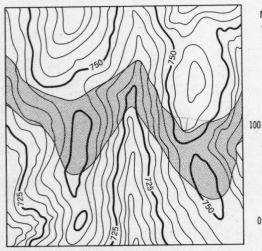

FIGURE X3.1

FIGURE X3.2

3. Three points are located on a single structural plane. Point B is 200 m from point A on a bearing of N 20 E, and is 65 m lower. Point C is 250 m from A, on a bearing of N 65 E, and is 45 m lower. What is the attitude of the plane?

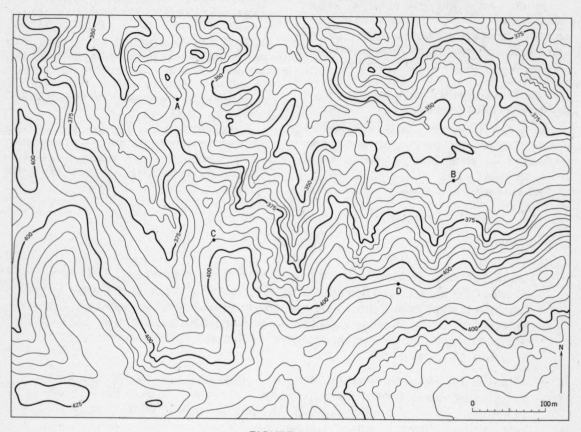

FIGURE X3.3

4
Lines and Intersecting Planes

DEFINITIONS

LINE
A geometric element that is generated by a moving point and has only extension along the path of the point. Lines may be rectilinear (straight) or curvilinear (curved). Only rectilinear elements are considered in this section.

PLUNGE
The vertical angle between a line and the horizontal (Fig. 4.1). Plunge is a special case of *inclination* and is analogous to the dip of a plane.

TREND
The strike of the vertical plane containing a line. Trend is included in the general term *bearing*.

PITCH
The angle, measured in some specified plane, between a line and the horizontal (Fig. 4.2). *Rake* is synonymous, but is not widely used.

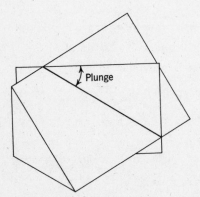

FIGURE 4.1 Plunge is measured in the vertical plane containing the line.

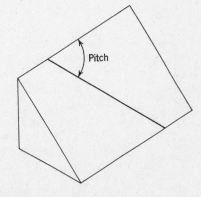

FIGURE 4.2 Pitch is measured in the inclined plane containing the line.

23

LINE

Lines of geologic interest may be marked by the long axes of mineral grains, streaks of minerals, slickensides on fault surfaces, striations, and by many other such features. Very elongate bodies of rock and drill holes may also be considered essentially linear for some purposes. An important class of lines are formed by the intersection of two planes.

PLUNGE

The trend and plunge uniquely define the attitude of any line in space. In referring to this attitude it is necessary only to quote the values of the two angles. The usual form is 30, S 45 E, which means that the line plunges 30° to the southeast. In the field the angles are measured in a manner similar to that used to obtain dip and strike (Fig. 1.3). As with dip and strike, lines are represented on a map by means of special symbols, also consisting of three parts:

1. A trend line
2. An arrowhead indicating the direction of plunge
3. A number giving the angle of plunge

The common symbols are shown in Fig. 4.3. Special symbols may be invented when needed, and all symbols used must be explained in the map legend.

For lines lying on planes, such as slickensides on fault surfaces or mineral lineation on foliation planes, the angle of plunge is, in effect, an apparent dip of the plane. The plunge angle is therefore confined to a range of values from zero to the true dip. The intersection of two planes is the line of apparent dip common to both planes.

PROBLEM

Determine the attitude of the line of intersection of the two planes N 21 W, 50 E, and N 48 E, 30 NW.

APPROACH

From the connection between plunge and apparent dip, it follows that the trend of the line of intersection lies between the two lines of true dip, and its angle of plunge is less than the dip of the more gently inclined plane. A down-dip view of the planes, aided with the use of two flattened hands, helps visualize these relationships. In this example, the trend must be northerly, and the plunge angle less than 30° (Fig. 4.4a).

CONSTRUCTION

1. Plot the strike lines of the two planes in map view. These are structure contours on the two planes which intersect at some point O (Fig. 4.4a).
2. To obtain another pair of structure contours
    (a) Folding about a line perpendicular to the trace of Plane 1 (FL 1) and using the angle of dip, locate point A at depth d on the plane.
    (b) Similarly, folding about FL 2 locate point B on Plane 2 using the same 'd'.

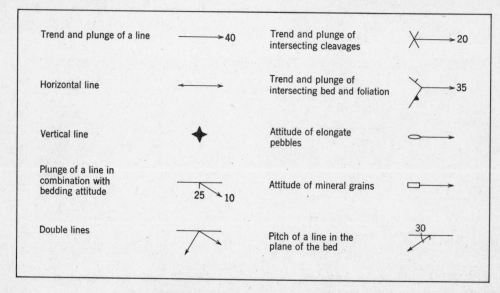

FIGURE 4.3 Map symbols for lines.

3. Project *AC* and *BC* parallel to the respective strike directions. *OC* represents the line of intersection in map view.

4. Using *OC* as a folding line, and the same depth *d* fixes point *D* on the line of intersection. Angle *COD* is the plunge.

ANSWER

The line of intersection trends due north, and plunges 24° in that direction.

PITCH

Plunge is measured in a vertical plane which is only incidently related to the structural plane in which the line lies. In contrast, pitch is measured on the plane, and is, therefore, a more direct description of the attitude of the line. The angle of pitch is always 90° or less. In describing pitch, it is necessary only to state the angle and identify the direction in which the acute angle faces. For example, a pitch of 35 N means that the angle faces in a northerly direction or alternatively, that it was measured downward from the north end

of strike line. In the field, the pitch of a line may be easily measured with a protractor if the plane surface is exposed. However, the exposure is usually not so convenient, and the measurement may be difficult. If pitch is required, it may be calculated from the plunge and dip or equivalent information.

PROBLEM

A line plunges 40, N 45 W on a plane with attitude N 90 E, 50 N. What is the pitch of the line on the plane?

APPROACH

As in the field, measurement of the pitch angle on a graphic construction requires that the plane be viewed normally.

CONSTRUCTION (Fig. 4.5a)

1. Plot the strike of the plane and the trend of the line in map view, intersecting at point *O*.

2. Establish FL 1 perpendicular to the line of strike. On a vertical section through FL 1 plot the dip angle and locate a point *A* on the inclined plane at some convenient depth *d* below the surface.

3. Project *A* back to the map view to fix the position of point *B* on trend line directly above plunge line, directly above *A* on the plane of

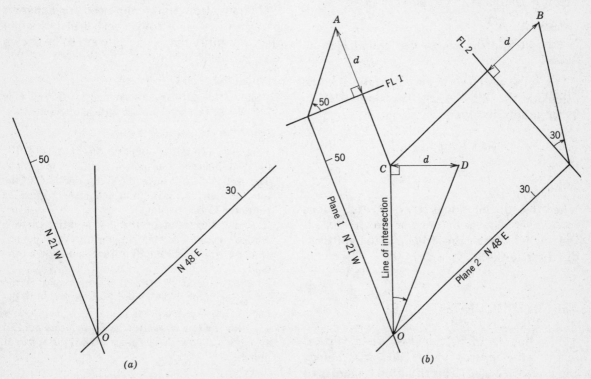

FIGURE 4.4 Plunge of the line of intersection of two planes. (*a*) Visualize the line with the aid of down-dip views. (*b*) Construction of the angle of plunge.

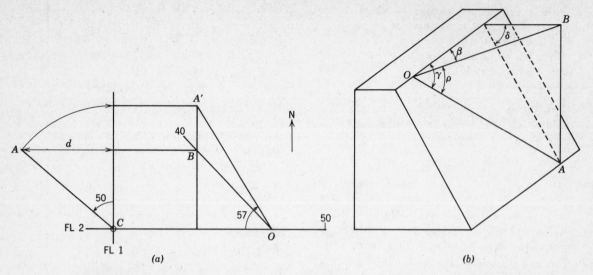

FIGURE 4.5 Pitch of a line in an inclined plane. (a) Construction of the angle of pitch. (b) Block diagram showing the various angles involved in the pitch calculation.

the map. Draw a line perpendicular to the strike through B and beyond.

4. Using the strike line as FL 2, rotate the inclined plane to horizontality. During this rotation A moves along the arcuate path whose center is point C. BA' is the projection of this path in the plane of the map. Therefore A'OC is the pitch angle.

ANSWER

The pitch of the line on the inclined plane is 57 W.

For some purposes it may be convenient to calculate the pitch angle trigonometrically. The relationships are

$$\sin r = \sin \beta \, / \cos \delta \qquad (4.1)$$

or

$$\cos r = \cos \beta \, \cos p \qquad (4.2)$$

Where r and p represent the pitch and plunge respectively, and as before δ is the dip and β is the angle between the strike and the apparent dip direction (see Fig. 4.5b).

APPARENT PLUNGE

If a line is to be depicted on a vertical section, the *apparent* rather than *true* plunge angle must be used. This situation is similar to the use of apparent dip on sections oblique to the true dip direction. In contrast to the relationship between apparent and true dip,

apparent plunge is always greater than true plunge. The maximum possible value of 90° occurs on sections which strike perpendicular to the trend of the line; the minimum value is equal to the true plunge and occurs on sections parallel to the trend of the line.

One situation where the need for apparent plunge arises is if nonvertical drill holes and the rock units they penetrate are to be shown on a vertical section.

PROBLEM

Show the attitude of an inclined drill hole (30, N 45 E) on an east-west vertical section.

CONSTRUCTION (Fig. 4.6)

1. Draw a line representing the trend of the plunging line and the line of section in map view. The section may include the surface point O of the line, as here, or not, in which case O must be projected to the section.

2. Using the trend of the line as FL 1, draw a vertical section showing the angle of true plunge, and at a convenient depth d locate point W on the line.

3. Project point W back to the map view of the line giving point X, which in turn is projected to the line of section giving point Y.

4. With the line of section as FL 2, locate point Z at depth d below Y. The angle YOZ is the apparent plunge.

ANSWER

The apparent plunge of the inclined drill hole on an east-west section is 39°.

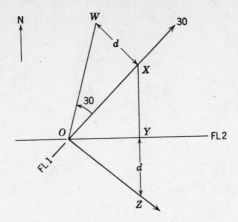

FIGURE 4.6 Apparent plunge.

EXERCISES

Using Fig. X4.1 (see p. X-2), structural plane *A* represents a shear zone with attitude N 66 E, 50 S, and plane *B* a limestone bed with attitude N 22 W, 40 W. Determine the orientation of the line of intersection of the two planes, the pitch of this line in plane *B*, the surface outcrop point of the line, and the depth at which the line would be found by drilling in the bed of Boulder Creek.

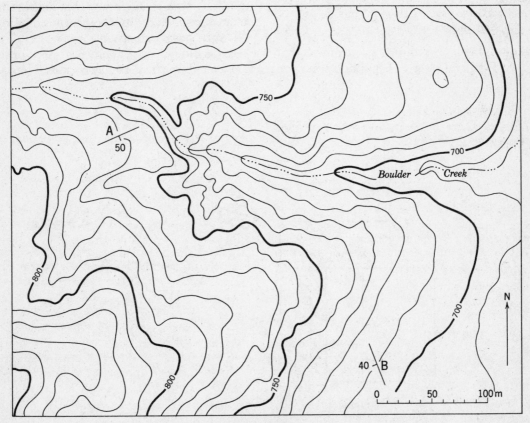

FIGURE X4.1

5
Concepts of Strain

INTRODUCTION

When a body is subjected to a system of forces it will undergo one or more of the following (see Fig. 5.1);

1. translation: transport relative to some coordinate system.
2. rotation: change in orientation.
3. distortion: change in shape.
4. dilation: change in volume.

It is convenient to group these displacements into two general classes. Rigid body motions are those which involve a change in the position of all the particles of a mass relative to a set of coordinate axes. This is the subject of dynamics, a branch of mechanics which studies the relationship between force and motion. Change in the positions of the particles relative to one another is termed *deformation*. The study of the deformed state is another, though far more complex branch of mechanics, and involves the relationship between stress and strain.

In geology, as in physics, the problems associated with these two classes are quite

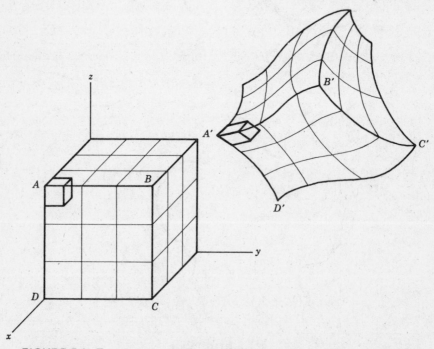

FIGURE 5.1 Translation, rotation, distortion and dilation of reference cube.

different. For example, determining the distance of transport of an essentially intact thrust sheet may involve working out the original sedimentary facies relationships in rocks of the same age found above and below the thrust plane. On the other hand, the study of folding, which may involve little or no transport, consists of making a comparison of the structure with the original, pre-deformation condition of the rocks.

Strain analysis is essentially a geometrical description of the deformed state. In even a simple structure this description may be quite complicated. Imagine a reference body of known size and shape imbedded in a rock mass (for example, the cube in Fig. 5.1). After deformation this body will have a quite different configuration. In the general case the result will be irregular; original planes become curviplanar, original lines curvilinear. The strain is termed *inhomogeneous*. One way to proceed is to introduce a simplification. Consider a much smaller part of the irregularly strained mass (the small cube at corner *A*, Fig. 5.1). In contrast to the entire mass, the deformed equivalent of this small reference body has a high degree of regularity; though distorted, the planes are still planar, lines still linear. This state of strain is *homogeneous*.

Accordingly, this method of attack involves seeking such small parts for which the strain is effectively homogeneous, and then to build up an overall pattern either by comparing a series of adjacent homogeneous parts, or by extrapolation based on continuity of structures related to strain. In order to do this however, it is necessary to develop an understanding of the geometry of homogeneous strain. Although it is ultimately necessary to treat strain problems in three-dimensions,

many situations can be approached from just two. The simplifying assumption that the strain can be fully represented in two-dimensions also serves as a way of introducing the subject. One way of exploring the geometry of two-dimensional, or *plane* strain is with the use of card–deck models.

CARD-DECK MODELS

The technique itself is ultra-simple. Various forms are drawn, or printed, on the edge of a reasonably thick deck of cards. The deck is then sheared. This can be most easily done by flexing the deck with one end held firmly together, and then releasing the deck with the other end held firmly. The shape of the deck in edge view, together with the printed forms are now distorted. Technically, strain involves a *continuous* alteration of form, whereas the "deformation" of the card-deck is simulated by a series of small slips, with each card remaining undeformed (Fig. 5.2). However, the thinner the cards the closer a continuous distortion is approached. When forms of about a centimeter or larger are printed on a deck of thin cards, the distortions are for all practical purposes continuous. Discarded IBM cards are ideally suited for this purpose. They are uniform in size, thin, but sufficiently stiff, and they are readily available in quantity from any computing facility. Care is needed in obtaining cards from a single run, otherwise some difficulty may be experienced in uniformly shearing the deck.

For quantitative experiments it will be found convenient to have an accurately square, three-sided tray to hold the cards in

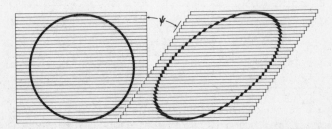

FIGURE 5.2 A card-deck with exaggerated card thickness, illustrating the formation of the strain ellipse by homogeneous simple shear. The angle ψ is the angle of shear.

the sheared position (see Ragan, 1969a, p. 135). The results can then be analyzed directly on the cards, or the patterns can be reproduced quite easily on a flat window copying machine.

HOMOGENEOUS STRAIN

This type of distortion of the card-deck is called simple shear, a *very* special type of strain. However, despite this special character, the models illustrate many properties of more general deformation, including the existence of a component of rotation. Strain is homogeneous if straight lines remain straight, and parallel lines remain parallel. In the card-deck models these conditions are met by insuring that the edge of the sheared deck remains straight; this is easy to do with the flexing mechanism. The simplest parameter of simple shear is the angle of shear (ψ), that is, the change of an original right angle. In the models this is easily measured by starting with square-ended deck (see Fig. 5.2).

The most basic experiment is to homogeneously deform a deck on which a circle has been drawn. An ellipse results—this is the strain ellipse (see Fig. 5.2 and 5.3). That such a transformation produces an exact ellipse can be demonstrated mathematically (Ramsay, 1967, p. 58). Although the emphasis in a card deck model is on the plane edge of the cards, the deformation is, of course, taking place throughout the deck, that is, in three dimen-

sions. The original circle can, therefore, be thought of as the diametral plane of a sphere embedded in the deck. As a result of deformation, this sphere becomes an ellipsoid; the ellipse on the edge of the card deck is a principal plane of this ellipsoid. After producing a pronounced, and reasonably large ellipse, the following observations and measurements can be made (Fig. 5.3):

1. Locate and draw in the two mutually perpendicular axes of the ellipse. Clearly, these were diameters of the original circle (and sphere), and represent the lines of maximum and minimum elongation with lengths $(1 + e_1)$ and $(1 + e_3)$. In simple shear, lengths parallel to the intermediate axis of the ellipsoid, that is, perpendicular to the plane edge of the card deck, remain unchanged $(1 + e_2 = 1.0)$. The values of the new lengths can be computed from the formula $e = (l_f - l_i)/l_i$ where l_i and l_f are the initial and final lengths respectively. Thus the extension e (which may be either positive or negative) is a simple parameter of change in length.
2. Similarly, the extension associated with any line of intermediate orientation can also be determined, for example, along the coordinate axes (e_x and e_y).
3. It is then possible to compute area change $\Delta = (1 + e_1)(1 + e_3) - 1$. For the card-deck ellipse this computation shows that $\Delta = 0$; there is no change in area in simple shear.

4. By superimposing a concentric circle on the final ellipse identical with the starting circle, two special lines are identified which have undergone no net change in length (*ab* and *cd* of Fig. 5.3). These two lines are called *lines of no finite longitudinal*

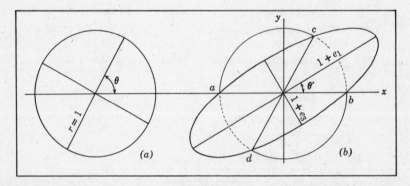

FIGURE 5.3 Some geometric properties of a unit circle transformed into an ellipse by simple shear. (*a*) The original circle with the mutually perpendicular lines destined to become the axes of the ellipse. The angle θ defines the orientation of these lines. (*b*) The strain ellipse. The angle θ' fixes the orientation in the ellipse axes.

strain. In simple shear one of these always coincides with the direction of shear.

5. Calculating the sums $[1/(1 + e_1)^2 + 1/(1 + e_3)^2]$ and $[1/(1 + e_x)^2 + 1/(1 + e_y)^2]$ demonstrates that they are equal, and this relationship holds for any two perpendicular directions. The constant value of this sum is called the *first strain invariant* (J_1). This also suggests that parameters of change of length in addition to e may be mathematically convenient. Accordingly, the quadratic elongation $\lambda = (1 + e)^2$, and the reciprocal quadratic elongation $\lambda' = 1/\lambda$ are defined. Despite the seemingly complex form of these new parameters, a simplification actually results from their use. In particular, various problems involving strained objects can be solved with them (see Chapter 6).

6. Returning the card-deck to the starting position, the lines which were the principal axes of the ellipse are also perpendicular in the starting circle. These two lines are unique. No other lines remain perpendicular after strain. These are termed the principal axes of strain. During simple shear, and generally, they rotate through angle ω ($= \theta - \theta'$; Fig. 5.3). This is the rotational component of strain. The whole deformation may then be conceived of as involving two components: a solid body rotation through angle ω, followed by pure (irrotational) strain, that is, by stretching and contracting in the direction of the respective ellipse axes without change in orientation.

7. Note, too, that on restoring the deck to the initial position, the circle used to define the lines of no finite elongation becomes an ellipse identical in shape to the strain ellipse but inclined in the opposite direction. This ellipse, which becomes a circle on deformation, is called the *reciprocal strain ellipse*. Its axes are also coincident with the mutually perpendicular lines in the starting circle.

PROGRESSIVE DEVELOPMENT OF THE STRAIN ELLIPSE

Important as these geometrical features of the strain are, it is even more important to gain some idea of how they develop and change progressively during the deformation. Initially this can be done rather rapidly, perhaps several times, by shearing the deck through a series of small increments. However, a much more precise picture can be obtained by plotting certain changes in length

and orientation against the increasing angle of shear. For example:

1. Plot the values of $(1 + e_1)$, $(1 + e_3)$ and $R_s = (1 + e_1)/(1 + e_3)$ (see Fig. 5.4a).

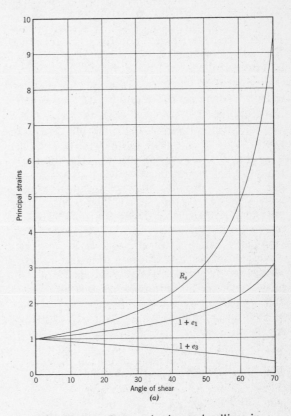

FIGURE 5.4 Changes in the strain ellipse in simple shear. (After Ramsay, 1967, p. 85–86.) (*a*) Values and ratio of the principal strains. (*b*) Orientation of the principal extension.

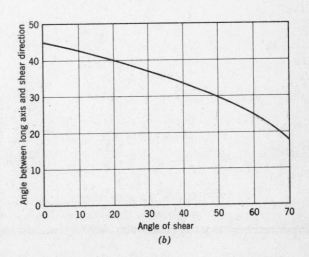

2. Similarly, plot the angle between the long axis of the ellipse and the direction of shear (see Fig. 5.4b). Some difficulty may be experienced in accurately locating the axes of the ellipse for small strains; it may be helpful to make the measurements as the deformation of the card deck is progressively removed. It is even more difficult to determine the orientation at the very start of the deformation; here it may help to realize that the lines of no longitudinal strain start at the 0° and 90° positions.

3. Plot the values $(1 + e_1)$ against $(1 + e_3)$ for increasing deformation (Fig. 5.5). This particular graph gives a history of the stages through which the ellipse passes in reaching some final state of strain. The curve depicting this history is called the *deformation path*, and for constant area ellipses it lies along the curve defined by the relationship $(1 + e_1) = 1/(1 + e_3)$. This type of two-dimensional deformation plot does not take into account the rotational component of the deformation. This is a rather severe limitation since it does not, for example, separate strain ellipses formed by pure shear and simple shear. However, while incomplete, it does clearly demonstrate that the results of the models are fully applicable to general two-dimensional, constant area strain, once the patterns of the final strain state are divorced from the card-deck.

4. The angle of rotation ω can be taken into account in a three-dimensional diagram (Fig. 5.6). With this graph it is clear that there are many other equal-area ellipses which do not form by the cases pure and simple shear.

SUPERIMPOSED STRAIN

The various graphs, especially the deformation plot, bring out clearly the fact that the strain ellipse passes through a whole series of intermediate ellipses in arriving at the final state of strain. This progression can be thought of in terms of a series of increments in which strain ellipse is superimposed on strain ellipse. Deforming a card-deck with both an ellipse and a circle illustrates the geometry of such superimposed strains (Fig. 5.7). In general, the orientation of the initial ellipse, the final ellipse, and the strain ellipse will not be the same. Further, the ratios of the axial lengths of the initial ellipse R_i, the final ellipse R_f, and the strain ellipse R_s are not simply related. These features are brought out most clearly by eliminating the component of solid body rotation (Fig. 5.8). A before and after comparison clearly demonstrated that material lines, such as those parallel to the initial ellipse axes, change orientation. There is also an apparent rotation of the final ellipse because the vertex occupies successively different material points during deformation.

It should be clear that the orientation ϕ_f and the shape R_f of the final ellipse depends on the angle which the initial ellipse makes with the maximum principal strain direction

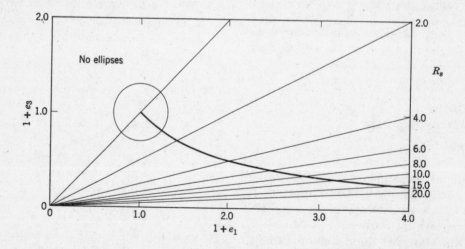

FIGURE 5.5 Two-dimensional deformation path. The single path shown is for constant area ellipses. If an area increase is involved in the deformation the ellipses plot above this curve; those involving a decrease below it. (After Ramsay, 1967, p. 95.)

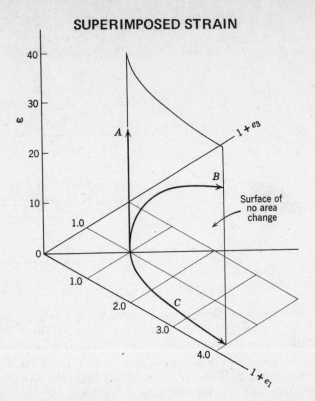

FIGURE 5.6 Three-dimensional deformation paths. Line A shows the path for rigid body rotation, line B for simple shear ellipses, and line C represents ellipses formed by pure shear (no rotation). All these paths lie on the surface of constant area. (After Ramsay, 1967 p. 96.)

ϕ_i and on the values of R_i and R_s. The mathematical aspects of these relationships have been discussed in detail by Ramsay (1967, p. 91-94, 204-211), Dunnet (1969),

and Elliott (1970), who also gives a graphical method for solving the equations. The main features can also be introduced with a card-deck experiment. If a series of diversely oriented constant-ratio ellipses are deformed,

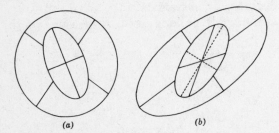

FIGURE 5.7 Superimposed strain ellipses. (a) An ellipse, representing a certain state of strain, and a larger reference circle drawn on the card-deck. (b) After deformation by simple shear. Note that the axes of the initial ellipse (solid lines), the axes of the now twice-deformed ellipse (dashed lines), and the principal axes of the additional strain do not coincide. (After Ramsay, 1967, p. 92.)

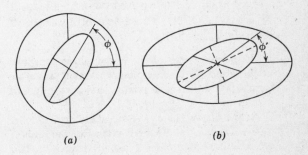

FIGURE 5.8 The same forms as in Fig. 5.7 adjusted to remove the rotational component of strain.

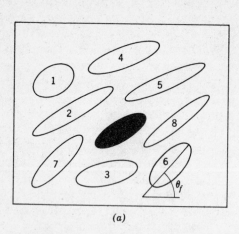

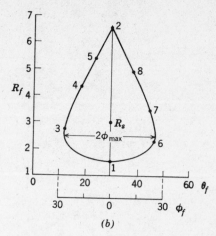

(a)   (b)

FIGURE 5.9 Homogeneous deformation of variously oriented, originally constant ratio ellipses. (a) The results of deformation with $R_s > R_i$. The angle θ_f fixes the orientation of the long axis of each elliptical form. The solid black ellipse is the strain ellipse (initially a circle). (b) A graph of R_f against the angle θ_f and ϕ_f for each ellipse illustrates the relationship between the deformed and strain ellipses. The ratio of the strain ellipse (R_s) plots near the center of the tear-drop curve. (After Ramsay, 1967, p. 211.)

the resulting forms will have certain characteristics. In the example (Fig. 5.9a) there is a diversity of shapes and tendency for the long axes to be aligned parallel to the direction of maximum elongation.

This pattern is best brought out by plotting R_f for each ellipse against an orientation angle θ_f measured from an arbitrary reference line (Fig. 5.9b). The resulting points lie on a closed curve symmetrical about a fixed value of θ_f, which defines the orientation of the long axis of the strain ellipse $\phi_f = 0$. Two points on this line represent the special cases in which the initial ellipse and the directions of the principal strains are coaxial (ellipses 1 and 2 in Fig. 5.9a). These two ellipses are of additional interest:

1. If the major axis of the initial ellipse is parallel to the direction of maximum extension, a narrow form is produced whose axial ratio is the product of the two ratios: $R_{f\max} = R_s R_i$.
2. If the major axis of the initial ellipse is parallel to the direction of minimum extension, the ratio of the resulting broad form is the quotient of the two ratios: $R_{f\min} = R_s/R_i$. (This is true only where $R_s > R_i$, as in the example. If $R_s < R_i$ a somewhat different approach must be used, as discussed below.)

Thus with $R_{f\max}$ and $R_{f\min}$ known from the plot, the values of R_s and R_i can be determined.

$$R_s = (R_{f\max}R_{f\min})^{1/2}$$
$$R_i = (R_{f\max}/R_{f\min})^{1/2}$$
(5.1)

This graph clearly brings out an additional feature of such deformed ellipses. There is a distinct and systematic variation in the orientation of their long axes. A measure of this

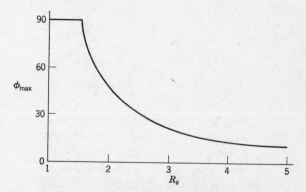

FIGURE 5.10 Typical curve showing variation of fluctuation $\phi_{\max}$ with increasing strain ratio R_s. (After Elliott, 1970, p. 2232.)

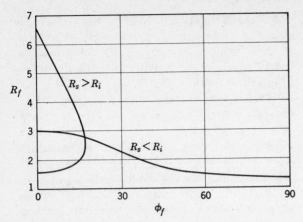

FIGURE 5.11 A R_f/ϕ plot showing curves for $R_s > R_i$ and $R_s < R_i$.

strain. For the original random orientation, and for the early stages of deformation ϕmax = 90°, that is the ellipses have no preferred orientation. The turning point occurs when $R_s = R_i$; then with a further small increment of strain a strong orientation "springs" into existence (Fig. 5.10). All the changes can be followed during the progressive shearing of a card-deck model.

When $R_s < R_i$, and ϕmax = 90°, the R_f/ϕ plot gives a curve with a broad peak (Fig. 5.11). Theoretically, R_s and R_i may still be determined (in this case $R_f$min = R_i/R_s), but practically, a scatter of points will make it difficult to estimate the orientation of the strain ellipse, and the values of the extreme ratios. Ramsay (1967, p. 210) discusses several other aspects of this problem.

variation is the angular difference between the long axes of the extreme orientations (*3* and *6* of Fig. 5.9). This angle is termed the *fluctuation* 2ϕmax and it is a function of the relative values of R_i and R_s. Where $R_s > R_i$ as in the example, the fluctuation is fairly small, and would continue to decrease with further

Once it is seen how these ellipses combine, it should be apparent that something similar must happen during each increment of progressive strain. This can be illustrated, though

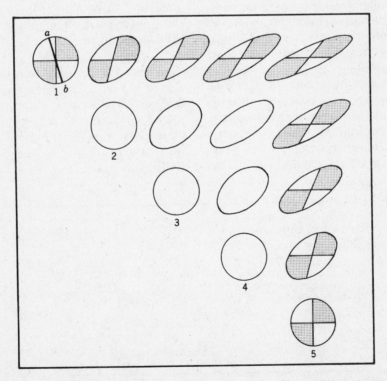

FIGURE 5.12 Finite increments in progressive deformation. Stippled sectors are zones of net extension at each increment. Row 1 gives the strain history. Column 5 depicts the changes occurring during the final increment. Note that the sectors do not coincide; some lines in the contracted zones must be actually extending.

crudely, by deforming a reference circle through a series of finite increments, and after each step adding a new reference circle to the deck. Following the final increment the various ellipses will then record in some detail what happened at each stage in the history of the development of the final state of strain as represented by the deformed equivalent of the initial circle (see Fig. 5.12). At every step of the way, there must exist lines of no longitudinal strain which divide the ellipses into two pairs of sectors. Material lines lying in the sectors which include the major axis experience elongation, whereas lines in the other sectors undergo shortening. As already noted, the lines of no longitudinal strain originate at the 0° and 90° positions at the very instant deformation begins.

Each new circle added records the changes occurring during subsequent steps. These additional figures also have associated zones of extension and contraction. Finally, following the last step (see column 5, Fig. 5.12) all the ellipses record what *must* have occurred during this last increment of strain. Observe, however, that the lines bounding the zones of extension and shortening for each final ellipse have different orientations. Clearly then some lines in the shortening zones experience an elongation, and similarly some lines in the extending zone experience a shortening at some stage in their history. This can be readily demonstrated with the card-deck model by focusing attention on the history of certain lines. For example, a material line lying in the

contraction field but close to the initial line of no longitudinal strain will initially undergo shortening (see line *ab*, Fig. 5.12). However, it will rotate more rapidly than the line of no longitudinal strain (not a material line), and thus will sweep through it into the zone of extension. Thereafter it will experience a progressive lengthening.

Since the development of the strain ellipse is a continuous process, an analysis based on a few finite increments can be only an approximation of what actually goes on during progressive strain. A more accurate picture could be obtained by considering a larger number of smaller increments, though there are obvious limits to doing this in practice. A completely accurate progression is obtained when, in the manner of differential calculus, the increments are considered to be infinitesimally small. This leads to the idea of the *infinitesimal* strain ellipse; that is, one which departs only slightly from the reference circle. Infinitesimal strains are commonly used in engineering to describe elastic behavior; this use involves simplifying assumptions which make them mathematically easier to work with, and they apply with sufficient accuracy when extensions are less than about 1 percent. The same approach is also useful in understanding progressive *finite*, that is, large strain which can now be conceived in terms of a series of infinitesimal increments superimposed on the developing finite strain ellipse. In simple shear the infinitesimal principal strain axes lie at angles of

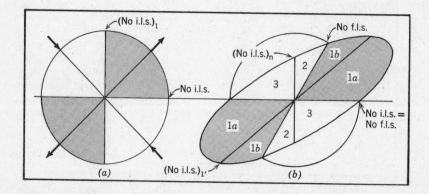

FIGURE 5.13 Arrangements of zones of elongation and contraction is simple shear.
(*a*) Infinitesimal strain ellipse (no i. l. s = lines of no infinitesimal longitudinal strain).
(*b*) Finite strain ellipse (no f. l. s = lines of no finite longitudinal strain).

45° to the shear direction, and the lines of no infinitesimal longitudinal strain are parallel and perpendicular to this direction (Fig. 5.13a). Each infinitesimal strain increment has exactly this same orientation. As the strain progresses, the material line coinciding with the initial line of no infinitesimal longitudinal strain (indicated by the subscript 1 in Fig. 5.13a) is rotated. Following the final increment this line lies within the extending zone (indicated by subscript 1' in Fig. 5.13b). The final infinitesimal increment also has associated with it a line of no infinitesimal longitudinal strain (indicated by the subscript n). The position of these lines, together with the lines of no finite longitudinal strain, define several zones each of which has a characteristic history of contraction or extension or both (Fig. 5.13b).

Zone 1 is the zone of finite extension. Lines in subzone *1a* have a history of continuous elongation; lines in subzone *1b* have an early stage of contraction (the line *ab* in Fig. 5.12 is an example).

Zone 2 contains lines which were contracted but experienced a late stage elongation. The net change is still one of shortening.

Zone 3 is the zone of continuous contraction.

Types of strain other than simple shear develop zonal arrangements with differing degrees of symmetry. Some understanding of these zones and their evolution is important for several reasons. First, they relate to visible structures which may be present in the rocks: thin competent layers oriented in Zone *1a* will undergo boudinage, while those in Zone *3* will be folded; layers in both Zone *1b* and *2* will show folding with a late stage of unfolding or boudinage, with the degree of late stretching greatest in Zone *1b*. Second, it illustrates vividly the inherent complexity of finite strain. Finally, while the final form of the strain ellipse is independent of the type of strain, the history of the zones of contraction and expansion are not, and potentially at least it may be possible to say something about the deformation path on this basis.

It should be emphasized that the positions of these zones also shift if area changes take place. This can be illustrated by comparing the model ellipse with circles either slightly larger or smaller than the starting circle. It will then also be appreciated that for large area changes, lines of no finite longitudinal strain may not exist at all. In such cases, all lines, irrespective of their orientation, will be either elongated or contracted.

INHOMOGENEOUS STRAIN

With an extension of the simple models used in illustrating the development of the strain ellipse, a good introduction to inhomogeneous strain can be obtained with a card-deck on which a large number of small circles have been printed. The deck is then deformed with a continuously varying angle of shear (Fig. 5.14a). While the resulting ellipses completely define the variation in strain, it is convenient to reduce this information to an even simpler form. This is done by constructing two sets of curves, called strain

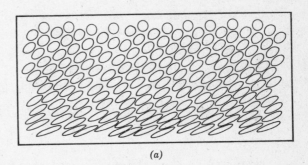

(a)

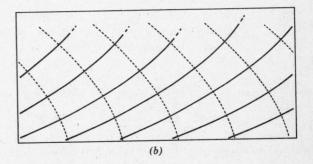

(b)

FIGURE 5.14 Inhomogeneous strain. (*a*) Field of strain ellipses. (*b*) Trajectories of the principal strains: $(1 + e_1)$ is solid, $(1 + e_3)$ is dashed.

trajectories, which are everywhere tangent to the ellipse axes (Fig. 5.14*b*). Note that the values of the strain vary along these curves, thus only orientational data is given directly. However, the $(1 + e_1)$ trajectories converge toward areas of greater strain, so that the patterns *do* indicate the variation in the values of the principal strains as well.

EXERCISES

1. A repetition of the various illustrative card-deck experiments is an excellent way to fix the basic concepts of strain in mind. These should include:
   a. Production of an ellipse from a circle, with observation of the principal axes, the lines of no finite elongation, the rotation of the axes during progressive shearing, the first strain invariant, and the reciprocal strain ellipse. The simple shear deformation path should also be plotted, both in two- and three-dimensions.
   b. Deformation of an original ellipse, and the relationship between the orientation of its axes and the principal axes of strain.
   c. The illustration of progressive strain, with special reference to the zones of extension and contraction. Graphically follow the history of a line, such as line *ab* in Fig. 5.12, by plotting its length against an increasing angle of shear.

2. The changes in geometry can also be followed by plotting the orientation of the principal axes, and the lines of no finite elongation on the ordinate against the angle of shear on the abscissa. It is also instructive to graph the orientation of the axes of an ellipse whose long axis is originally parallel to the direction of shear.

3. Experiment with inhomogenous strain by deforming a large number of small circles (these can easily be stamped on the deck with a slipon pencil eraser and an ink pad).

4. Experimentally deform other shapes, including right triangles in a variety of orientations. For several of these right angles, measure the angle of shear.

5. For the two sets of deformed ellipses Fig. X5.1 and X5.2 (see p. X-3), determine the orientation of $(1 + e_1)$ and the values of R_i, R_s and ϕ_{max}. Discuss the results.

6. Consider the problem of the determination of state of strain in the case where the initial ellipses have a preferred orientation. Experimentally deform: (1) a group of perfectly aligned ellipses of uniform R_i, (2) a group of perfectly aligned ellipses of variable R_i. Discuss the results.

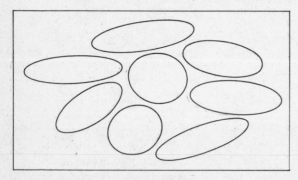

FIGURE X5.1

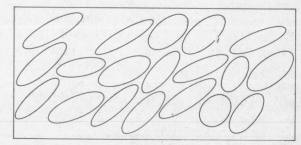

FIGURE X5.2

6
Analysis of Strain in Rocks

INTRODUCTION

As the card-deck models clearly demonstrate, most lines change length and orientation as a result of strain. Several different parameters of change in length are used: simple extension e is one of these; others include quadratic elongation λ, and reciprocal quadratic elongation λ'. Generally, if two lines were initially perpendicular, the original 90° angle will also change during strain. A measure of this change is the angle of shear ψ (see Fig. 5.2); a mathematically more convenient parameter is the *shear strain* $\gamma = \tan \psi$.

With suitable material, one or more of these parameters can be determined by measuring objects of known original size or shape preserved in the deformed rocks. From these measurements, certain aspects of the irrotational component of strain can be deduced. For example, measurement of the shape of originally spherical oolites yield the orientation and magnitudes of the principal strains directly. The same information can be obtained from diversely oriented, originally ellipsoidal pebbles as discussed in Chapter 5. This section is devoted to additional examples from which two-dimensional strain information can be obtained, and to the methods used. A word of caution: the strain deduced from such objects will depart from the strain

in the embedding material according the degree of contrast in the mechanical properties. The assumption made in the examples given here is that this contrast is negligible; that is, that both object and matrix deform homogeneously. In practice this may not be valid, and must be tested.

SIMPLE STRAIN ANALYSIS

Fossils often possess such features as planes of symmetry, known angular relationships, or proportions which are constant in individuals of a given species. They are, therefore, common objects of known original shape.

PROBLEM
Given a collection of deformed brachiopods (Fig. 6.1a), determine as much about the strain ellipse as possible. Note that in the undeformed state, the hinge and the trace of the symmetry plane are perpendicular.

CONSTRUCTION (After Wellman, 1962)
1. Transfer the hinge and the symmetry line of each individual fossils to a tracing sheet (Fig. 6.1b).
2. On this tracing locate a line of arbitrary length and orientation. It is convenient if this line is not parallel to any fossil line and at least several inches long (see line AB of Fig. 6.1b).
3. For each individual shell, draw lines parallel to

the hinge and symmetry lines through both A and B giving a parallelogram (see example drawn for shell no. 8).

4. Through all the pairs of fossil points determined in this way, and through points A and B, draw a best fit ellipse, either by sketching, by projecting a circle from an inclined position, or most satisfactorily with one of several special ellipse drawing devices which are commercially available.

5. Measure of orientation of the principal axes, and their lengths.

ANSWER

Since the size of the ellipse depends on the length of the chosen lines AB the absolute lengths of the axes can have no meaning; only their ratio can be determined: $(1 + e_1)/(1 + e_2) = R_S = 1.76$; and the $(1 + e_1)$ axis makes an angle of $10°$ with the hinge line of fossil no. 5.

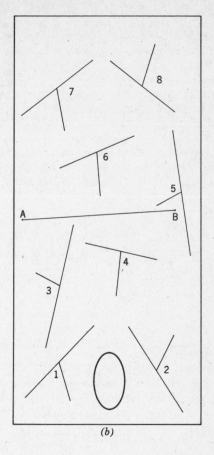

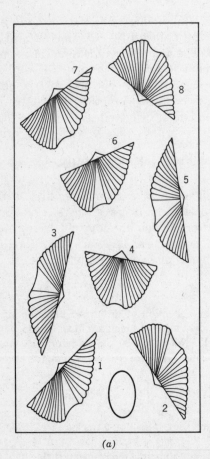

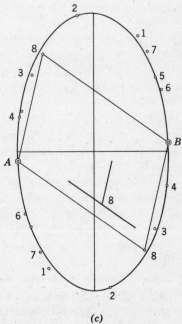

FIGURE 6.1 Simple graphical method for determining strain from deformed fossils. (a) Slab of deformed brachiopods. (b) Traces of hinge lines and lines of symmetry. (c) Strain ellipse.

APPROACH

In order to see why this method works, imagine having made the same construction before deformation. Because each pair of fossil lines was originally perpendicular, rectangles rather than parallelograms would have resulted. Collectively, the corners of all these rectangles would have defined a circle with AB as diameter. It is this circle from which the constructed strain ellipse is derived.

Now re-examine the deformed fossils. The deformed shape of each shell is a function of orientation. Strictly, all right angles have been eliminated. However, shell no. 3 is still nearly symmetrical. This is also the longest, and narrowest form. Shell no. 4 also retains an approximately 90° angle, but is deformed in an opposite way to a short, broad form. As demonstrated with the card-deck models, there are only two lines which are perpendicular before and after deformation—the principal axes of strain. Therefore the direction of maximum elongation nearly coincides with the hinge line of no. 3 and the symmetry line of no. 4, and the direction of minimum elongation is perpendicular to it. Thus one can estimate the orientation of the principal strain axes by inspection.

EQUATIONS DESCRIBING THE PROPERTIES OF THE STRAIN ELLIPSE

The strain ellipse found graphically from the deformed brachiopods by Wellman's method can also be determined analytically. To do this, expressions for the longitudinal and shear strain are required. The unit circle which becomes the strain ellipse is given by (Fig. 6.2a):

$$x^2 + y^2 = 1$$

The orientation of any line of radius OP can be defined by the angle which the line makes with the principal axis of strain $(1 + e_1)$. After strain, the resulting ellipse is (Fig. 6.2b):

$$\frac{x^2}{(1+e_1)^2} + \frac{y^2}{(1+e_2)^2} = 1$$

As a result of this strain line $OP = 1$ changes in length to become $OP' = 1 + e$, with its orientation now defined by ϕ'. Also after strain, the tangent at point P' is no longer perpendicular to OP', the acute angle between the normal to this tangent and OP' is the angle of shear ψ associated with the line.

From these geometrical features we seek to derive equations for longitudinal and shear strain in terms of the principal strains and the orientation. The following treatment is adapted from Ramsay's excellent book (1969, especially p. 65-70), which should be consulted for further details and examples.

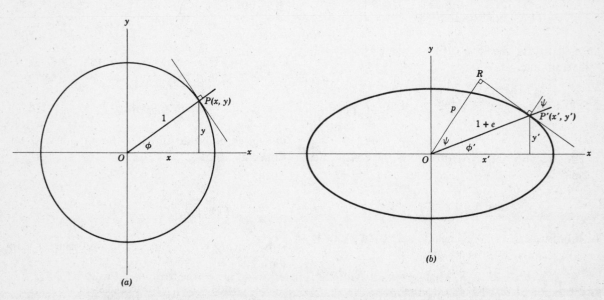

FIGURE 6.2 Geometry of the strain ellipse: (a) unit circle, (b) derived ellipse.

Changes in Length. The stretched length of any line is the product of the original length and the radius of the strain ellipse parallel to it. This applies not only to $OP' = (1)(1 + e)$, but also to its components x' and y'. Therefore

$$x' = x(1+e_1) \quad \text{and} \quad y' = y(1+e_2) \qquad (6.1)$$

Also, from Fig. 6.2a

$$x = \cos\phi \quad \text{and} \quad y = \sin\phi \qquad (6.2)$$

Substituting (6.2) into (6.1)

$$x' = (1+e_1)\cos\phi \quad \text{and} \quad y' = (1+e_2)\sin\phi$$

Then, by the Pythagorean theorem

$$(1+e)^2 = (1+e_1)^2 \cos^2\phi + (1+e_2)^2 \sin^2\phi$$

Rewriting, using quadratic elongations $\lambda = (1 + e)^2$

$$\lambda = \lambda_1 \cos^2\phi + \lambda_2 \sin^2\phi \qquad (6.3)$$

With this, the extension of any line defined by ϕ can be calculated, knowing the principal strains. However, when dealing only with deformed material, the angle of interest is ϕ', not ϕ. From Fig 6.2b

$$x' = (1+e)\cos\phi' \quad \text{and} \quad y' = (1+e)\sin\phi'$$

Substituting these into (6.1), using (6.2), and rearranging

$$\cos\phi = \frac{(1+e)\cos\phi'}{(1+e_1)} \quad \text{and} \quad \sin\phi = \frac{(1+e)\sin\phi'}{(1+e_2)}$$
$$(6.4)$$

The trigonometric relation linking $\cos\phi$ and $\sin\phi$ is

$$\cos^2\phi + \sin^2\phi = 1$$

Substituting (6.4), again using quadratic elongations, and rearranging

$$\frac{1}{\lambda} = \frac{\cos^2\phi'}{\lambda_1} + \frac{\sin^2\phi'}{\lambda_2}$$

Or, using reciprocal quadratic elongations $\lambda' = 1/\lambda$

$$\lambda' = \lambda'_1 \cos^2\phi' + \lambda'_2 \sin^2\phi' \qquad (6.5)$$

With this we have the equivalent of (6.3) expressed in terms of the angles measured in the strain state, as required.

In addition to changes to length, the orientation of the line also changes. From Fig. 6.2b

$$\tan\phi' = y'/x'$$

Substituting (6.1)

$$\tan\phi' = y(1+e_2)/x(1+e_1)$$

From Fig. 6.2a

$$\tan\phi = y/x$$

Making use of this

$$\tan\phi' = \left[(1+e_2)/(1+e_1)\right]\tan\phi$$

or

$$\tan\phi' = (1/R_s)\tan\phi$$
$$(6.6)$$

Shear strain. From Fig 6.2b, where the perpendicular distance to the tangent line is $OR = p$

$$\sec\psi = (1+e)/p$$

Using the trigonometric identity $\tan^2\psi = \sec^2\psi - 1$, and the definition of $\gamma = \tan\psi$

$$\gamma^2 = (1+e)^2/p^2 - 1 \qquad (6.7)$$

The general equation for the tangent to an ellipse at point (x',y') is

$$\frac{xx'}{(1+e_1)^2} + \frac{yy'}{(1+e_2)^2} = 1$$

Substituting (6.1) and (6.2), and rearranging slightly

$$\frac{\cos\phi}{(1+e_1)}x + \frac{\sin\phi}{(1+e_2)}y = 1 \qquad (6.8)$$

This is the tangent line in terms of the principal strains and the orientation of the line in question. Point R is the simultaneous solution of this tangent line and the line normal to it. Writing the equations for these two lines in abbreviated form

$$ax + by = 1$$
$$bx + ay = 0$$

where a and b represent the coefficients of x and y in (6.8). By squaring both, adding and rearranging

$$x^2 + y^2 = 1/(a^2 + b^2)$$

But $p = (x^2 + y^2)^{1/2}$, therefore, in the specific terms of the problem

$$p = \left[\frac{1}{(\cos^2\phi)/(1+e_1)^2 + (\sin^2\phi)/(1+e_2)^2} \right]^{\frac{1}{2}} \quad (6.9)$$

Substituting this expression for p into (6.7), making use of $\lambda = (1 + e)^2$, and rearranging

$$\gamma^2 = \lambda(\cos^2\phi/\lambda_1 + \sin^2\phi/\lambda_2) - 1$$

Substituting the expression λ from (6.3)

$$\gamma^2 = (\lambda_1 \cos^2\phi + \lambda_2 \sin^2\phi)$$
$$(\cos^2\phi/\lambda_1 + \sin^2\phi/\lambda_2) - 1 \quad (6.10)$$

Expanded, this is a fourth degree equation of seemingly complex form. However, if $(\cos^2\phi + \sin^2\phi)^2 = 1$ is similarly expanded and subtracted from (6.10), then on combining, rearranging terms and taking the square root a considerable simplification results.

$$\gamma = \frac{\lambda_1 - \lambda_2}{(\lambda_1 \lambda_2)^{1/2}} \sin\phi \cos\phi$$

Again the need is for γ expressed in terms of ϕ'. Substituting (6.4), and simplifying

$$\frac{\gamma}{\lambda} = \frac{1}{\lambda_2} - \frac{1}{\lambda_1} \cos\phi' \sin\phi'$$

Using $\lambda' = 1/\lambda$, and for further convenience defining $\gamma' = \gamma/\lambda$

$$\gamma' = (\lambda_2' - \lambda_1') \cos\phi' \sin\phi' \quad (6.11)$$

With (6.5) and (6.11) we can calculate both the elongation and the shear strain for any line. For example, if $(1 + e_1) = \sqrt{2}$, $\lambda_1 = 2.0$, and $\lambda' = 0.5$, and $(1 + e_2) = \sqrt{1/2}$, $\lambda_2 = 0.5$, $\lambda_2' = 2.0$ (note that $R_s = 2.0$), then a line defined by $\phi' = 20°$ has the following associated strain components.

From (6.5)

$$\lambda' = 0.5 (.940)^2 + 2.0 (.342)^2 = 0.68$$

$$(1+e) = \sqrt{1/\lambda'} = 1.22$$

And from (6.11)

$$\gamma' = (2.0 - 0.5) (.940(.342)) = 0.48$$
$$\gamma' = \gamma/\lambda \text{ or } \gamma'/\lambda' = 0.73$$
$$\psi = 35°$$

These values can be verified by measurements on Fig. 6.2b.

MOHR'S CIRCLE FOR FINITE STRAIN

These equations can be arranged in a still more useful form. Substituting the double angle identities

$$2\cos^2\phi = 1 + \cos 2\phi$$
$$2\sin^2\phi = 1 - \cos 2\phi$$
$$2\sin\phi\cos\phi = \sin 2\phi$$

into (6.5) and (6.11), and rearranging gives

$$\lambda' = \frac{\lambda_1' + \lambda_2'}{2} - \frac{\lambda_2' - \lambda_1'}{2} \cos 2\phi'$$

$$\gamma' = \frac{\lambda_2' - \lambda_1'}{2} \sin 2\phi' \qquad (6.12)$$

These now have the form of the parametric equations

$$x = c - r \cos\alpha$$
$$y = r \sin\alpha$$

which describe a circle centered at $(c,0)$ with radius r in an xy-coordinate system. This property of (6.12) allows problems involving

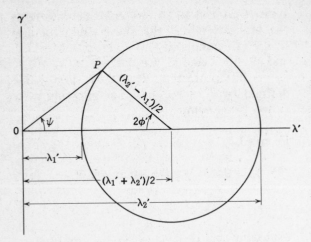

FIGURE 6.3 Mohr's circle for finite strain.

λ' and γ' to be solved graphically with a *Mohr's circle* construction for finite strain (Fig. 6.3). The main features of this plot are a circle with a center located on the abscissa (λ'-axis) a distance of $(\lambda_1' + \lambda_2')/2$ from the origin and a radius of $(\lambda_2' - \lambda_1')/2$. As an example, the previous numerical problem can be solved in this manner. The center of the circle is located (2.50/2) units from the origin, and a circle (1.50/2) units in radius is drawn (again see Fig. 6.3). The line defined by $\phi' = 20°$ must be plotted as $2\phi' = 40°$ and measured

from the long axis of the strain ellipse, that is from λ_1'. The coordinates of the point P on the circumference so located are the required values of λ' and γ'. Note that the angle of shear ψ can be obtained directly by this method, thus bypassing the mathematically convenient, but otherwise obscure parameter γ'.

The most important use of the Mohr's circle construction is not, however, to solve for the values of λ' and γ', but to determine the size or shape of the strain ellipse and its orientation from measurements made on deformed material. To illustrate the type of problem that can be solved with a Mohr's circle, the analysis of the deformed brachiopods will be repeated, two of which are shown in Fig. 6.4.

PROBLEM

Given deformed symmetric shells, or any original right angles, determine the shape and orientation of the strain ellipse.

APPROACH

In order to determine the strain, enough information is required to draw the circle. For angular measurements only this means that the angle of shear for each of two forms, and their relative orientation must be determined; additional measurements serve as a check.

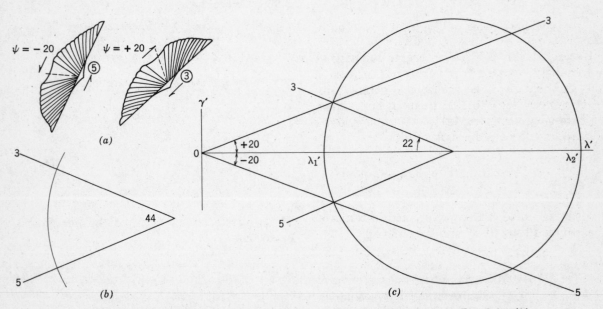

FIGURE 6.4 Analysis of deformed fossils. (*a*) Two brachiopods from Fig. 6.1*a*. (*b*) Relative orientation of the hinge lines using twice the angle. (*c*) Mohr's circle constructed from the two shells.

CONSTRUCTION (see also Ramsay 1967, p. 237)

1. Measure the angle of shear for each shell, with attention to the sign of the shear. In Fig. 6.4a, the measurements are made relative to the hinge line. If the symmetry line is used as datum, the angles will be numerically the same, but the signs reversed. Either will work, so long as the measurements are consistent.

2. Measure the relative orientation of the hinges of the shells: in the example, the angle between no. 3 and no. 5 is 22°.

3. On a sheet of tracing paper. plot the relative orientations of the hinges from a common point, using twice the measured angle. Using any convenient radius, draw a circular arc using the point of intersection of the lines as center (see Fig. 6.4b).

4. On a second sheet, draw the two coordinate axes and plot the two angles of shear from the origin, measuring from the λ'-axis. Positive angles are plotted above the line, negative below.

5. Superimpose the tracing with the hinge orientations and the arc on the coordinate axes. By moving the center of the arc along the λ' - axis, and at the same time rotating the sheet, find the position at which the hinge lines and their corresponding angles of shear lie on the arc. This gives the center of the Mohr circle, which can now be drawn in on the diagram.

ANSWER

The ratio of the principal strains, found by measuring the λ'_1 and λ'_2 intercepts using any convenient scale is $R_s = (\lambda_2'/\lambda_1')^{1/2} = 1.75$. This is in good agreement with the results of the Wellman method, especially considering that the strain is not strictly homogeneous (see slight scatter of points in Fig. 6.1), and that small plotting errors are inevitable. Note that the method gives the same orientational information for an arc of any radius (see step 3). As before, only the ratio of the values of the principal strains can be found. The principal directions can also be determined. The angle between λ_1' and the line representing shell no. 3 is $2\phi' = 22°$. Therefore the direction of maximum elongation is $\phi' = 11°$ measured from the hinge line of shell no. 3 away from no. 5 on the original slab. This also checks with the results of the Wellman construction.

Situations arise where angles other than right angles are deformed; these too can be used to give strain information, provided the value of the original angle is known.

PROBLEM

Given the deformed glass shards typically found in welded tuffs, determine the strain associated with their formation (Fig. 6.5a). Assume that the shards originally form at the junctions of three bubbles, and thus had angles of 120° separating the limbs.

APPROACH

Basically, the method is the same as applied to original right angles. An intermediate step is required, however, in order to convert the measured angular changes into ψ.

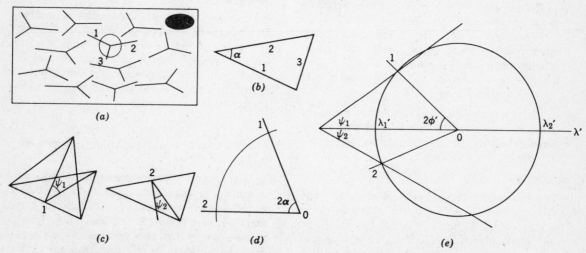

FIGURE 6.5 Strain from deformed glass shards in welded tuff. (a) Simulated shards. (After Ragan and Sheridan, 1972.) Circled form is analyzed for strain. (b) Equivalent scalene triangle. (c) Construction of angle of shear for non-right angles. (d) Orientation of shard limbs using twice the angle. (e) Mohr's circle constructed from the deformed shard.

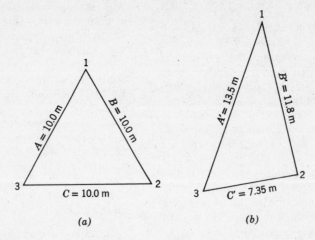

FIGURE 6.6 Stakes set in the surface of a glacier in the form of a triangle: (a) initial configuration, (b) deformed equivalent.

CONSTRUCTION (after Ramsay, 1967, p. 245)

1. The shard with prongs 1, 2, and 3 is equivalent to a scalene triangle with sides 1, 2, and 3 (Fig. 6.5b).

2. Originally, this triangle was equilateral. Comparing the figures before and after deformation, the angle of shear for each side can be constructed. In Fig. 6.5c, the perpendicular bisector of side 1 originally connected the opposite corner, and now connects the same corner of the deformed triangle. This gives the angle of shear associated with side 1 directly (ψ_1). This same construction is repeated for side 2, giving ψ_2.

3. The complete Mohr's circle can now be drawn exactly as before:

    a. The angle between sides 1 and 2 is a'; this is plotted $2a'$ on tracing paper, together with an arc of arbitrary radius (Fig. 6.5d).

    b. The angles of shear (ψ_1 and ψ_2) are plotted on the coordinate axes.

    c. Superimposing the tracing on the λ' - axis, the position where the intersections of the corresponding lines lie on the arc establishes the center, and the full circle is drawn (Fig. 6.5e).

ANSWER

The results from the complete Mohr's circle gives a strain ellipse with R_S = 2.0, and with long axis in the plane of the horizontal foliation (compare with strain ellipse shown in Fig. 6.5e). If only angular measurements are made, no absolute values of the extensions can be determined, and therefore neither can area changes be calculated. However, in the case of welded tuffs, the change in porosity independently demonstrates that the deformation is compactional in nature. The measured strain there-

fore indicates an area (volume) reduction of 50%, and thus the long axis of the strain ellipse is also the line of no finite elongation.

PROBLEM

Three stakes are set in the surface of a glacier so as to form the corners of an equilateral triangle 10 m on a side (Fig. 6.6a). After one year this triangle is resurveyed, and found to be distorted as shown in Fig. 6.6b. Determine the magnitude of the principal strains and their orientation.

APPROACH

The three original known lengths are equivalent to a rosette of strain gauges (Fig. 6.7a). Thus after strain (Fig. 6.7b) the elongation along these lines can be deduced. This is sufficient information to obtain a solution. Graphically, the only problem is to find the circle which will satisfy the measured strains and their orientation.

CONSTRUCTION (adapted from Holister, 1967, p. 62-66)

1. Rearrange the line elements into a sequence of ascending or descending magnitudes, so that the included angle between the maximum and minimum observed strains is less than 180°. Here this requires only that the position of the line of intermediate strain be adjusted. Note the angles in the rearranged configuration (Fig. 6.7c).

2. For each line compute the reciprocal quadratic elongation. Label the lines A, B, and C, so that $(1 + e_a) \rangle (1 + e_b) \rangle (1 + e_c)$. In this example, $\lambda_a' = 0.55$, $\lambda_b' = 0.71$, $\lambda_c' = 1.85$.

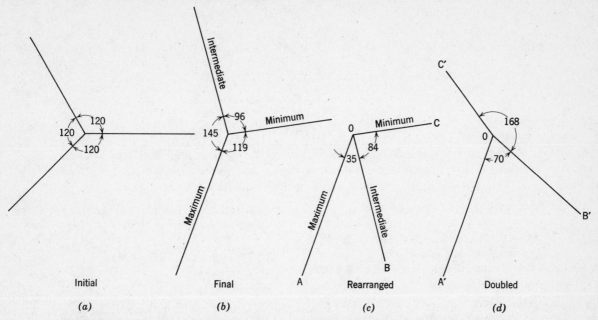

FIGURE 6.7 Rosette of strain gauges: (a) initial configuration, (b) deformed equivalent, (c) rearranged strain lines, (d) relative orientation angles doubled.

3. From any vertical datum, which serves as the γ'-axis, mark off three parallel lines distances a, b, and c to the right of the datum line which are proportional to the computed elongations λ_a', λ_b', λ_c', using any convenient scale (Fig. 6.8).

4. From any point P on the intermediate strain line (b) draw straight lines PQ and PR, making angle AOB and BOC are obtained from the rearranged axes.

5. Draw a circle through point P, Q, and R. The center can be located at the intersection of the perpendicular bisectors of PQ and PR. This gives

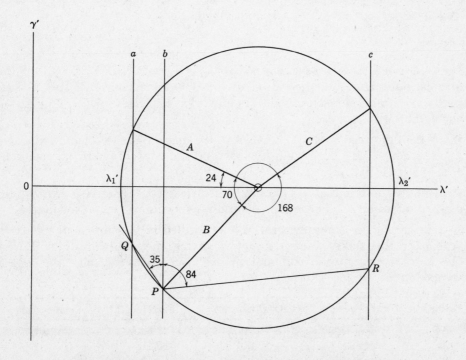

FIGURE 6.8 Mohr's circle for the surface ice of the glacier.

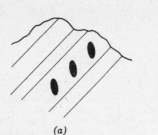   

| (a) | (b) | (c) | (d) |

FIGURE 6.9 Rotation component of strain. (After Ramsay, 1969, p. 54.) (a) Strained and rotated bedded rocks. (b) Relationship of bedding to strain ellipse. (c) Original orientation of bedding to strain ellipse. (d) Bedding restored to horizontality.

the Mohr's circle for the observed strain field, the ordinate being the γ' - axis and the abscissa being the λ' - axis drawn to pass through the center of the circle. The principal strains are represented by the intercepts with the abscissa, read using the same scale.

6. Replot the rearranged line elements so that the angles AOB and BOC are doubled (from Fig. 6.7d). These define the relative orientations of the elements on the $\lambda'\gamma'$ - plane. Generally the Mohr's circle cuts each of the strain lines a, b, and c twice. Superimposing the double angle diagram with 0 at the center of the circle and rotating until A corresponds with a, B with b, and C with c, identifies which of these intersections to draw in.

ANSWER

The principal reciprocal elongations are $\lambda_1' = 0.48$, $\lambda_2' = 1.98$. The angle between λ_1' and A is $2\phi' = 24°$. This defines the orientation of the long

axis of the strain ellipse $\phi' = 12°$. measured from A toward B in the physical plane (see Fig. 6.7b).

ROTATION COMPONENT OF STRAIN

If the pre-strain orientation of a line or plane is known, the rotation component of strain can be determined by a two step calculation. Removing the irrotational component changes the orientational angle from ϕ' to ϕ, according to the relationship given by (6.6). The line is then rotated through angle ω to restore it to the original orientation. One situation where this is possible is if sedimentary bedding is present, and it is assumed to have been originally horizontal (see Fig. 6.9a-d).

EXERCISES

1. Draw a Mohr's circle for finite strain for $(1 + e_1) = 2.3$, $(1 + e_2) = 0.7$. From the diagram determine the orientation of the lines of maximum angular shear, the lines of no finite elongation, and the extension associated with a line making an angle of 25° with $(1 + e_1)$.

2. Using the deformed fossils (Fig. X6.1, p. 49) determine the orientation and shape of the strain ellipse using Wellman's method and check the result by constructing a Mohr's circle diagram.

3. Analyze the deformation of the simulated welded tuff (Fig. X6.2). Assuming the strain to be wholly compactional, determine the volume change involved.

4. A triangle of three stakes was set in holes drilled in the surface of a glacier. One year later the stakes were resurveyed, giving the following data. Determine $(1 + e_1)$, $(1 + e_2)$, their orientation, the area change, and the rotational component.

Side	Initial bearing	Original length	Final length	Final bearing
AC	due north	17.00 m	21.61 m	N 15 E
AB	N 60 E	17.00	18.16	
BC	N 60 W	17.00	13.88	

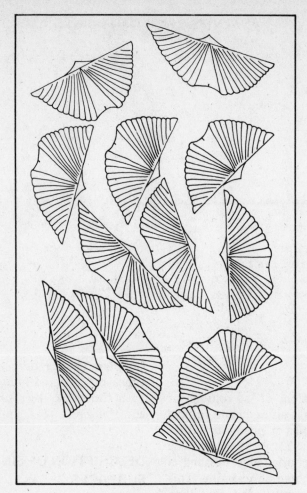

FIGURE X6.1

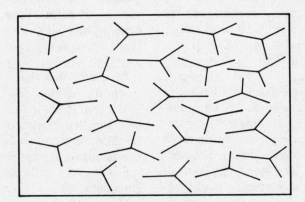

FIGURE X6.2

7

Description and Classification of Folds

INTRODUCTION

A *fold* is a distortion of a volume of material that manifests itself as a bend or nest of bends in linear or planar elements within the material (Hansen, 1971, p. 8). Most folds involve elements that originally defined a plane. Bedding is the common example. This is an important case because the fold then represents an important indicator of the distortion, and its geometric features can be correlated with certain aspects of local strain, rotation, translation, and so on (Turner and Weiss, 1963, p. 105). However, folds may also develop from originally curved elements. *Folding* occurs when pre-existing elements are acted upon in such a way as to be transformed in curviplanar or curvilinear configurations, whatever their original condition. It is worth noting that the deformation that produces a fold in one situation may not in another. Initially curved elements might become planar or linear, the elements may be oriented so as to remain planar or linear (Ramsay, 1967, p. 473), or such marking elements may be absent from the rock mass entirely. In short, there may be "folding" without folds.

In the following sections, several simple geometric properties of folded planes are explored; the methods and terminology follow closely Fleuty (1964) and Ramsay (1967,

Chapter 7). For details of features and fold types not covered here these works, together with the books by Turner and Weiss (1963) and Hansen (1971), should be consulted.

DESCRIPTION OF SINGLE SURFACES

Single curviplanar surfaces may have a wide variety of forms, ranging from comparatively simple, such as shown in Fig. 7.1, to exceedingly complex. The geometry of even a relatively simple curved surface may be quite difficult to describe. There are mathematical methods of defining the geometry of any surface, but these have found little practical use. For irregularly curved surfaces alternative methods are required; one such method is to use structure contours (see Chapter 18).

Fortunately, it is meaningful to restrict our consideration to a much simpler class of surfaces. Many natural folded surfaces have shapes which closely approach the form of cylinders, or are made up of approximately cylindrical parts. A cylindrical surface is defined as one which can be generated by a line moved parallel to itself in space. The orientation of this generating line is a directional property of the entire surface, and has no particular location. It is analogous to a crystallographic axis, and is called the *fold*

FIGURE 7.1 A single curviplanar surface. The fold form in the front is approximately cylindrical.

axis. One of the most important geometric features of cylindrical surfaces is that their shape can be fully represented in a cross-section drawn perpendicular to the axial direction; this section is called the *profile* of the cylindrically folded surface.

Each profile curve has several geometrical features which serve to identify certain locations on the surfaces. The *crest*, or high point and the *trough*, or low point, on the curves are two such features. In three-dimensions each of these is actually a line on the cylindrical surface parallel to the axis. The location of both of these is dependent on the orientation of the surface relative to hori-zontal. On the other hand, the point of maximum curvature, or *hinge* point, and the point where the curve changes from concave to convex, or *inflection* point, are independent of orientation and thus serve to define the geometry of the surface more fundamentally (Fig. 7.2).

Often, single hinge and inflection points alternate. In three dimensions such points lie on hinge lines and inflection lines, and it is convenient to consider a single fold as the portion between two inflection lines. If a portion of the surface has the form of a circular arc, a specific hinge point, or hinge line, does not exist; in such instances it is

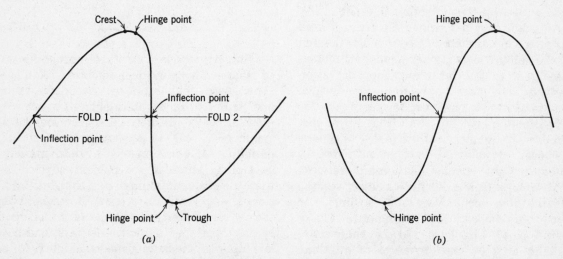

FIGURE 7.2 Identifiable points on the profile of a cylindrical surface.

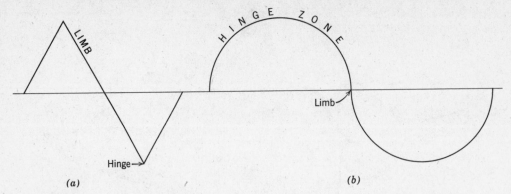

FIGURE 7.3 Differing proportions of fold surface made up of hinge zone and limb.

arbitrarily identified as the bisector of the circular segment. Similarly, there will be no inflection point if the transition from concave to convex involves a straight segment; the inflection point is taken to be the mid-point of this straight section.

The terms *hinge zone* and *fold limb* and the distinction between them have been precisely defined in a form which is useful for some purposes (Ramsay, 1967, p. 345). Here, however, a more general meaning is adopted which follows conventional usage (Dennis, 1967, p. 88, 102). The hinge zone is considered to be that portion of the curved surface adjacent to the hinge line, and the fold limb that part of the surface adjacent to the inflection line. The proportion of the entire curved surface which may be considered to be hinge zone or limb may vary. The extreme examples are: (1) if the hinge zone is reduced to a line, and (2) if the limb is represented by the inflection line (Fig. 7.3).

An additional descriptive element of fold shape is the *interlimb* angle, which is defined as the minimum angle between the limbs as measured in the profile plane; or, alternatively, the angle between the lines which are tangent to the profile curve at the inflection points (Fig. 7.4). This angle describes the "tightness" of the fold. For general purposes, however, it is often sufficient to categorize the angular relationship between fold limbs with the descriptive adjectives; the terms gentle, open, close, tight, isoclinal, and mushroom-shaped are common. Fleuty (1964, p. 470) has suggested that these terms be restricted to specific ranges of interlimb angles (see Table 7.1).

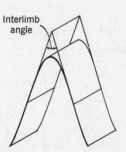

FIGURE 7.4 Interlimb angle.

TABLE 7.1 Terms Describing the Tightness of a Fold (after Fleuty, 1964, p. 470)

Interlimb angle	Description of fold
180° – 120°	Gentle
120° – 70°	Open
70° – 30°	Close
30° – 0°	Tight
0°	Isoclinal
Negative angles	Mushroom

Symmetry, or lack of it, is another feature of shape of cylindrical surfaces. A single symmetrical fold is one whose profile shape and its mirror image are identical. A series of linked folds is symmetrical if each member is symmetric, and if the pattern is strictly periodic. A consequence is that the two enveloping surfaces are planar and parallel, the surface containing all the inflection points, or *median* surface is mid-way between the two enveloping surfaces, and the plane of symmetry is perpendicular to these surfaces and to the median surface. These features make it easy to describe the dimensions of the

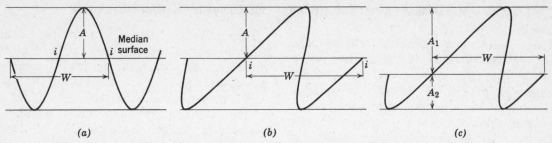

(a)   *(b)*   *(c)*

FIGURE 7.5 Dimensions of symmetric and asymmetric folds: W = wavelength, A = amplitude.

folds in terms of amplitude and wavelength (see Fig. 7.5a). Fold shape is related, in part, to the ratio of these two measures.

For asymmetric folds a description of size becomes increasingly involved as the degree of asymmetry increases (Fig. 7.5b,c). Complete schemes to describe the dimensions of such folds have been suggested (Fleuty, 1964, p. 467; Ramsay, 1967, p. 351; Hansen, 1971, p. 9).

RELATIONS BETWEEN ADJACENT SURFACES

Since folds almost invariably involve more than one surface, additional terms and methods are needed to establish the spatial and geometric relationships between adjacent curviplanar surfaces of the fold. The locus of all the hinge lines is one important feature, especially from the point of view of field mapping. The surface so defined is often referred to as the axial plane or surface, but because it is not directly related to the axis, this discrete surface is more appropriately called the *hinge surface* (Fig. 7.6). Like axis, *axial plane* should be reserved for the planar direction parallel to the hinge surface throughout the fold, as in the phrase *axial plane* foliation. Similarly, there is an inflection surface, the locus of the inflection lines

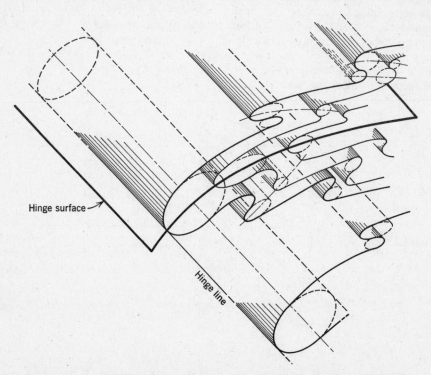

FIGURE 7.6 The hinge surface of cylindrical folds. (After Wilson, 1961.)

on successive surfaces, and crestal and trough surfaces.

The geometrical relationship between adjacent surfaces depends on the relative curvature of the two surfaces and the distance between them. The simplest and most sensitive way of defining this relationship is to construct lines of equal dip or apparent dip on the profile plane between the two surfaces; these lines are called *isogons* (Elliott, 1965; Ramsay, 1967, p. 363). Not only can the resulting patterns aid in distinguishing accurately between different fold forms, but their use also leads to a natural classification of fold geometry which has important bearing on fundamental questions of fold mechanics.

CONSTRUCTION (Fig. 7.7)

1. Obtain a profile of the folds to be analyzed. The most direct and accurate method is to trace the various curved surfaces from a photograph taken in the direction of the fold axis. If such a view is not possible, either because of lack of exposure or large scale, the profile view may be constructed from field data (see Chapter 10).
2. On the trace of each folded surface, draw a series of tangent lines using the horizontal as datum. A ten degree interval is convenient, but the spacing should be dictated by the actual fold form and the amount of detail required in each situation.
3. Connect points of equal dip on the two adjacent surfaces with a straight line—these are the isogons. Repeat for all layers of concern.

ISOGONAL CLASSIFICATION

Generally the isogons will not be parallel, and the degree of departure together with the direction of convergence (or divergence) establishes the basis of classification (Ramsay, 1967, p. 365). For consistency, the inner arc of the fold is taken as the reference point for statements about the direction of isogon convergence. Five types of patterns are recognizable, including three general and two special cases:

1. *Folds with strongly convergent isogons*: The curvature of the outer surface is less than that of the inner, and the smallest distance between the two surfaces occurs at the hinge (Fig. 7.8*a*).
2. *Parallel folds*: In this special case, the inner surface has a greater curvature than the outer surface, but their relationship is such that the isogons are perpendicular to these surfaces (Fig. 7.8*b*). Also the distance between the curves is everywhere constant; this distance is the orthogonal thickness of the folded layer.
3. *Folds with weakly convergent isogons*: The curvature of the inner surface is still greater, but the spacing between the bounding curves is greatest at the hinge (Fig. 7.8*c*).
4. *Similar folds*: Both curves are identical and the isogons are parallel. In this special case the distance between the curves measured along the isogons in constant, a measure known as the axial plane thickness (Fig. 7.8*d*).
5. *Folds with divergent isogons*: The curvature of the inner arc is less than that of the outer arc (Fig. 7.8*e*).

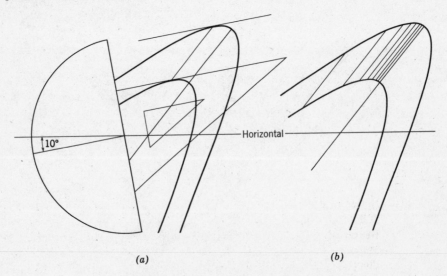

(a)   *(b)*

FIGURE 7.7 Construction of dip isogons.

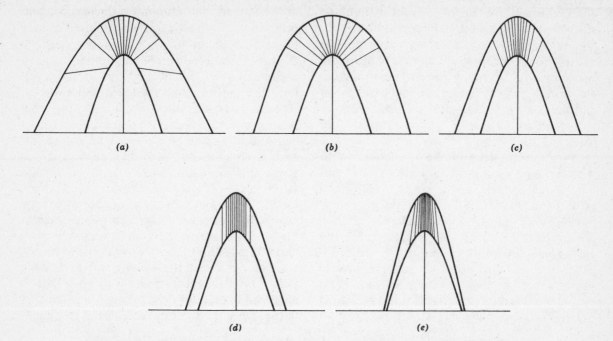

FIGURE 7.8 Classification of fold shape based on isogon pattern: (*a*) strongly convergent, (*b*) parallel, (*c*) weakly convergent, (*d*) similar, (*e*) divergent. (After Ramsay, 1967, p. 365.).

FOLD ORIENTATION

The orientation of a fold is completely defined by the direction of closure, and the attitude of the hinge line and hinge surface.

There are three terms which are used to describe the direction of closure. An *antiform* is a fold which closes upward, and a *synform* closes downward. A fold which closes sideways is a *neutral* fold, or, defined more strictly, it is a fold with a hinge pitching at between 80° and 90° on the hinge surface. *Anticline* and *syncline* are reserved for folds with older and younger rocks, respectively, in their cores. Many anticlines are also anti-

forms, but it is possible for an anticline to be in synformal position, that is, upside-down. Such a fold is described as an synformal anticline.

The angles of dip and plunge fix the attitude of the hinge surface and the hinge line with respect to horizontal, and are also the basis for a classification nomenclature. In an effort to standardize usage and to increase precision, Fleuty (1964) suggested precise limits to a series of traditional terms for both dip and plunge (see Table 7.2). These terms can then be combined to describe fold attitude, for example, a steeply inclined gently plunging fold. Note however that not all

TABLE 7.2 Terms Describing the Attitude of Folds (after Fleuty, 1964, p. 483, 486)

Angle	Term	Dip of hinge surface	Plunge of hinge line
0°	Horizontal	Recumbent fold	Horizontal fold
1° – 10°	Sub-horizontal		
10° – 30°	Gentle	Gently inclined fold	Gently plunging fold
30° – 60°	Moderate	Moderately inclined fold	Moderately plunging fold
60° – 80°	Steep	Steeply inclined fold	Steeply plunging fold
80° – 89°	Sub-vertical	Upright fold	Vertical fold
90°	Vertical		

combinations are valid; a gently inclined steeply plunging fold is an impossibility.

It is useful to give additional attention to neutral folds. *Recumbent* applies to horizontal neutral folds; both hinge and hinge surface are horizontal. With *vertical* folds the hinge and hinge surface are vertical. Both terms are incorporated in Table 7.2. A neutral fold intermediate between these two extremes is *reclined*; more strictly, this describes a fold with a hinge surface dip of between 10° and approximately 80° and a hinge which has a pitch of more than 80° on the hinge surface. Since the dip can be greater than 80° and the plunge still less than 80° required for the vertical designation (e.g. dip = 82°, plunge = 79°), it is not practical to place a precise upper limit on the hinge surface dip. This minor discrepancy results from the use of plunge for some folds and pitch for others.

The flaw in this approach is that the hinge is referred to a vertical plane which usually bears no relation to fold geometry. A classification based solely on dip and pitch could be constructed to avoid this, but would itself have drawbacks, the gravest of which is that pitch is often difficult to measure in the field. A simpler and still more precise classification can be achieved by combining both the plunge and pitch schemes. A special triangular diagram involving the three variables is used (Rickard, 1971; see Chaudhuri, 1972, for a closely related approach). Dip and pitch pairs are plotted at the intersection of the straight lines whose scales are found along the base and left side (Fig. 7.9a). For plunge, the scale on the right and the curves are used. Each point uniquely represents a fold attitude and all possible attitudes can be represented (Fig. 7.10). Only the limiting orientation catagories

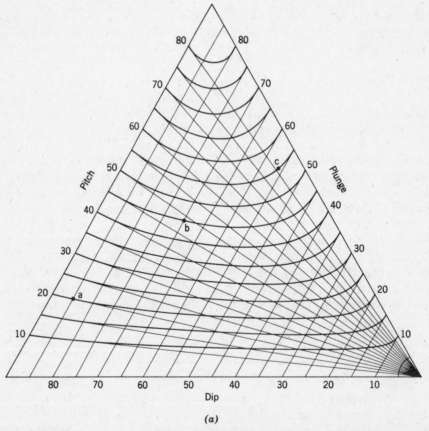

(a)

FIGURE 7.9a Fold attitude. Triangular grid for plotting fold attitudes.

with respect to the horizontal and vertical are needed. The fields of these catagories are shown in Fig. 7.9*b*. Recumbent folds overlap the reclined and horizontal folds, and vertical folds the reclined and upright fields (there is an additional small area of overlap where the fold name is optional). There is no special term for simple horizontal upright folds. Instead of further subdividing the large inclined field, Rickard usefully proposes that all folds be allocated an attitude index number according to their position in the diagram; dip *D* and plunge *P* are sufficient. The precise attitude of any fold can then be expressed by a simple descriptive name followed by the index in parentheses. The following examples will illustrate the method (each fold is plotted in Fig. 7.9*a*):

a. Upright fold ($D_{85}P_{20}$). According to the terminology of Table 7.2 this would be an upright gently plunging fold.

b. Inclined fold ($D_{70}P_{45}$). This is a steeply inclined moderately plunging fold.

c. Reclined fold ($D_{56}P_{55}$), or a moderately inclined reclined fold.

The diagram could also be used to bring out additional details concerning the folds of an area. If fold attitude changed in some geographic direction the plot of a number of points would bring out the change of the component of the attitude relative to the horizontal, or if an element of fold type, such as interlimb angle changed with attitude a series of plots would emphasize this relationship.

For completeness, the terms describing the direction of closure are then added, for example, upright synform or inclined antiform.

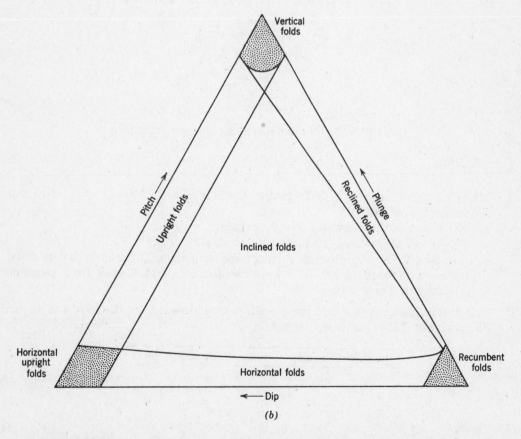

FIGURE 7.9b Use of the grid in classification of fold attitude. (After Rickard, 1971.)

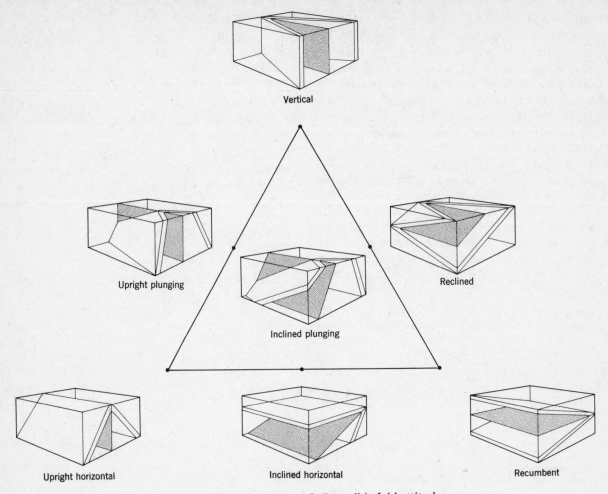

FIGURE 7.10   Illustration of all possible fold attitudes.

EXERCISES

1. Using the profile of the folded strata depicted in Fig. X7.1 (see p. X-4), construct the following features:

    a. The traces of the crest and trough surfaces.

    b. The traces of the hinge and inflection surfaces.

    c. Dip isogons at 20° intervals through the structure. Do certain layers show distinctive isogon patterns? Describe these patterns and discuss their meaning with respect to the structure.

2. Plot the following hinge and hinge surface combinations on Richard's triangular diagram (Fig. 7.9*a*), and name each fold.

	Hinge surface (dip)	Hinge line (plunge *p*, pitch *r*)
a.	85	*p* = 60
b.	80	*r* = 80
c.	65	*p* = 60
d.	83	*p* = 85
e.	20	*r* = 25

FIGURE X7.1

8
Parallel Folds

Parallel folds can be modeled by the simple bending of a card deck (Fig. 8.1). The total deformation is accomplished by the bending of the individual cards, and slippage at the surface of each card is an inevitable consequence of the process. The importance of this slip can be readily appreciated by trying to bend a deck which has been firmly clamped together at each end of the deck. By preventing the slip, the layered structure is effectively eliminated and the resulting deck is nearly as rigid as a block of wood. Clear evidence of this bedding-plane slip may be found in naturally occurring folds (Fig. 8.2).

Also, as a result of the bending, the surface of each layer is stretched along the outer arc and contracted along the inner arc. This in turn results in a thinning of the material adjacent to the outer arc and a complimentary thickening adjacent to the inner arc. These two regions are separated by a surface of no strain, called the *finite neutral surface*. While these changes of thickness tend to cancel, in general there will be a net change in thickness of the layer. Further, certain small-scale tensional and compressional structures may develop in these two regions of the layer. However, if the layers are thin as in the card deck model, the effect is slight and the change in thickness is negligible; therefore, the total thickness of the pack will remain effectively constant. (This can be confirmed by measurement of the bent card deck; of course, the cards must remain in contact during the bending.) With orthogonal thickness constant throughout, the model fold will be parallel.

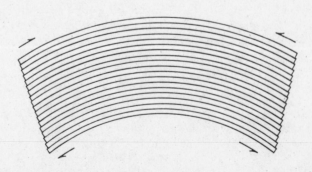

FIGURE 8.1 Card deck model of a parallel fold.

60

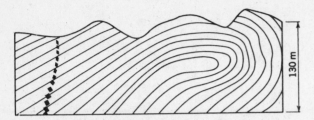

FIGURE 8.2 Basalt dike in the upper limb of an overturned fold cut by bedding plane faults. (See Capps, 1919, p. 41-42.)

To the extent that the mechanical properties of a rock mass undergoing folding approach those of a card deck, the natural folds will also have parallel form.

The physical nature of the cards also imposes a further condition on the three-dimensional shape of this simple model fold. It has a rectilinear hinge line, that is, the fold is cylindrical. Once the deck is folded, a second oblique bending by the bedding-plane slip mechanism is not possible. This is the property which gives corrugated sheet metal its rigidity. If the hinge line is curved it must be produced by some additional mechanism

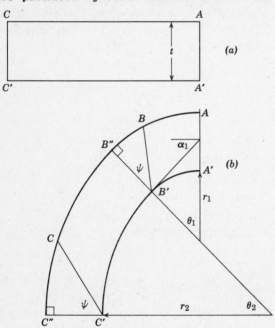

involving distortions within the plane of the layering. It should be noted here that with a more involved kind of bending it is possible to produce conical folds with a card deck. Although these may be important where cylindrical folds die out, they will not be discussed further here (see Wilson, 1967).

The bedding-plane slip mechanism is closely related to simple shear, except that the shear plane is curved and the angle of shear varies in the shear direction. The angle of shear is closely related to the fold shape, and can be easily calculated (Ramsay, 1967, p. 392). Suppose a fold is made up of two circular arcs, each defined by a radius of curvature and an angle (r_1, θ_1 and r_2, θ_2; Fig. 8.3). At point B', the dip $\alpha_1 = \theta_1$. The length of an arc is the product of the radius of curvature and the subtended angle expressed in radians. Therefore,

$$BB'' = AB'' - A'B' = \theta_1 t$$

Similarly, at C' the dip $\alpha_2 = \theta_1 + \theta_2 t$ and

$$CC'' = AC'' - A'C' = (\theta_1 + \theta_2)t$$

The amount of slip therefore depends on the angle of arc relative to the trace of the hinge surface, and is independent of the changes in curvature. The slip is also directly proportional to the thickness of the layers. The shearing strain can be expressed in terms of the angle of dip.

$$\gamma = \tan \psi = \alpha$$

This relationship has been solved for values of dip ranging from 0-90°, and the results are presently graphically in Fig. 8.4. In combination with Fig. 5.4a and 5.4b it is then possible to determine the values and orientation of the principal strains at any point in the folded layer.

PARALLEL FOLDS IN CROSS SECTION

The property of constant orthogonal thickness implies that a line perpendicular to one layer is also perpendicular to the layers above and below. This, and the fact that any curve

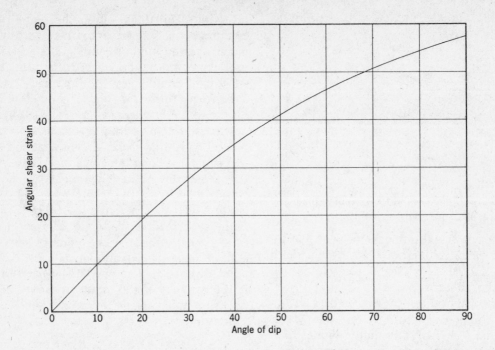

FIGURE 8.4   Angle of shear as a function of dip. (After Ramsay 1967, p. 393.)

can be expressed by a series of tangential circular arcs, form the geometrical basis for reconstructing parallel folds in cross section. This construction depends on the elementary proposition that the centers of two tangent circles lie on the straight line which passes through the point of contact and which is perpendicular to the common tangent (Busk, 1929, p. 13). In Fig. 8.5, two circles with centers at O_1 and O_2 touch at point A. Both O_1A and O_2A are perpendicular to the tangent AB, and therefore must lie on the same straight line.

The simplest application of this principle is to the problem of the shape of a curved layer

between just two successive dip measurements. As in Fig. 8.6, for each of the two dips A and B, draw normals OA and OB, intersecting at point O. With OA and OB as radii, draw the arcs AC and BD, which mark the form of the curve taken up by the layer. Note that the thickness of the bed exposed between A and B is $AD = BC$.

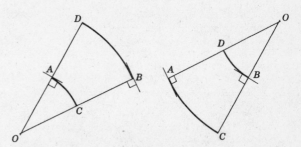

FIGURE 8.6   Tangent arcs through two adjacent dips. (After Busk, 1929, p. 14.)

This construction is easily extended to the case of three or more dip readings (Fig. 8.7). If A, B, and C are the dips, construct normals to the first pair A and B, locating the center of curvature O_1. Repeat for the second pair B and C, fixing center O_2. With appropriate radii, the tangent arcs may then be carried

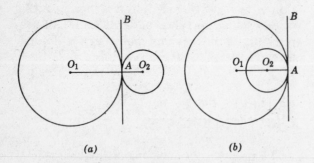

FIGURE 8.5   Tangent circles — the geometric basis for reconstruction of parallel folds. (After Busk, 1929, p. 13.)

across the structure from one dip normal to the next.

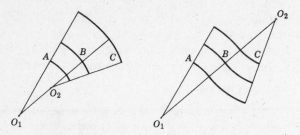

FIGURE 8.7 Tangent arcs through three adjacent dips. (After Busk, 1929, p. 16.)

If the attitude at two adjacent locations is the same, the normals constructed at the two dips *A* and *B* will be parallel and the required "arc" will be a straight line (Fig. 8.8). If the difference between successive measured attitudes is slight, the normals will intersect at some distance above or below the section, and the required radius may exceed the expansion of even a beam compass. This may be controlled by working at a smaller scale, or the large radius arc may be approximated.

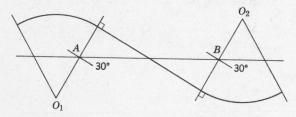

FIGURE 8.8 Reconstruction where adjacent dips are equal.

PROBLEM

Given the dip normals *AB* and *CD*, not far from parallel, draw a circular arc through point *A* (Fig. 8.9).

CONSTRUCTION (after Busk, 1929, p. 21)

1. Draw *AC* perpendicular to *AB*, and *AE* perpendicular to *CD*.
2. Bisect the angle *CAE*, with the bisector meeting *CD* at *G*.
3. Construct *GF* perpendicular to *CD*.
4. Point *G* is the intersection of the required arc with *CD*, and the arc may be easily sketched from *A* to *G*. The equal tangents *AF* and *FG*, may aid in positioning this arc.

The validity of this construction lies in the fact that the triangles *AOF* and *GOF* are equal, and therefore *OA = OG* (Busk, 1929, p. 22).

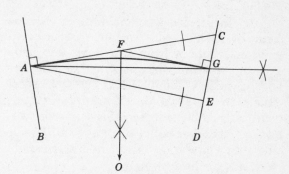

FIGURE 8.9 Reconstruction where the difference between adjacent dips is small.

The reconstructed fold must, of course, be in accordance with the evidence on the ground. If a recognizable horizon appears repeated along the line of traverse and the exposures are incomplete, there may be a small discrepancy between the actual and reconstructed positions. If one can confidently assume that the error is not due to other factors, such as thinning or faulting, an intermediate dip may be interpolated to adjust the actual and predicted position of the marker horizon. This dip may be interpolated anywhere along the traverse, but in the absence of any further information, it is probably best positioned where the control is least, that is, where the difference between adjacent dip angles is large, or the distance between measurements is great, or both.

PROBLEM

Given two adjacent dip readings, construct a curve to pass through both (Fig. 8.10).

CONSTRUCTION (after Higgins, 1962)

1. Label the lesser dip *A* and the greater dip *B*. Construct the normals to each dip intersecting at *C* and extending well beyond. Ordinarily *C* would be used as the center of curvature for the arc between these two normals.
2. Construct the perpendicular bisector of *AB*, which intersects the normal *AC* at point *Z*.
3. Arbitrarily locate point O_a on the normal *AC* as far beyond *Z* as is convenient.
4. Find *D* on the normal *BC*, so that $BD = AO_a$.

5. Find O_b at the intersection of the perpendicular bisector of line DO_a with BD.

6. O_a and O_b are the centers of two tangent arcs that define the required curve. The line O_aO_b is normal to the interpolated dip.

With these construction techniques, the form of folded beds may be reconstructed from map data or from a field traverse made expressly for the purpose. In either case, the line of section should be as nearly perpendicular to the strike direction of the dipping beds as possible. In preparing the cross section, it is conventional to orient the line so that its eastern or northern end is on the right hand side of the drawing.

Even in simple, regular folds, it is rarely possible to locate a section line exactly perpendicualr to all measured strikes, in which case apparent dips must be computed before they are plotted on the cross section. The correct elevations of the data points must, of course, also be used in plotting the dips.

A sufficient number of readings on the chosen section line is also rare. Other measurements may be used by projecting them short distances to the line of the section. Generally they are projected parallel to the line of strike, but this may require ad-justment for significant departures of the section line and dip directions. The need to convert to apparent dips and the use of obliquely projected dips means that folds are not truly cylindrical, and this introduces errors and uncertainities which must be kept in mind when interperting the reconstruction.

PROBLEM (Fig. 8.11)

Given the following dip angles corrected or projected to an east-west line as required, reconstruct the folds.

A = 20 E
B = 10 W
C = 45 W
D = 10 W
E = horizontal
F = 25 E
G = 75 E
H = 50 E
I = 20 W

CONSTRUCTION (after Busk, 1929, p. 19)

1. Draw normals to the measured dip at each outcrop point. These intersect successively at points O_1 through O_8.

2. With O_1 as center and O_1A as radius, draw an arc to the next normal O_1B thus establishing point K.

3. With O_2 as center and O_2K as radius, draw an arc to O_2C, giving point L.

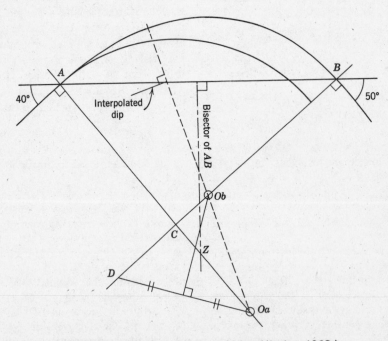

FIGURE 8.10   Interpolation. (After Higgins, 1962.)

4. Repeat using successive centers O_3 through O_8, giving arcs *LM, MN, NP, PQ, QR, and RS.* The curve *ALMNPQRS* is then the trace in cross section of the folded surface passing through point *A*.

5. This identical process may then be repeated to reconstruct the shape of the deeper horizons passing through points A_1 and A_2. For the even deeper horizons $A_3 - A_6$, a somewhat different approach is required. In order to maintain uniform layers, their thicknesses must be marked off along the normals O_8S or O_7R, and bringing these arcs into the core of the anticline produces angular rather than rounded hinges. This result is due to the fact that the centers O_5 and O_6 lie above horizon A_3, and thus can not affect the shape of the curves below them. At even deeper levels O_3 and O_4 are also eliminated.

6. The trace of the hinge surface may be drawn as a smooth curve through the sharp angular hinges in the deeper part of the anticline, and then extended as the bisector of arc *PQ*.

BALANCED CROSS SECTIONS AND THE DEPTH OF FOLDING

During folding *mass* is certainly conserved. Since density before and after deformation is essentially the same, volume must also be conserved (Goguel, 1962, p. 147). The simple bending of originally horizontal sedimentary layers, as with a card deck model, is an example of plane strain, and therefore the changes which occur can be fully depicted on the profile plane. If volume is conserved, then on this plane the area must also be conserved, and if bed thickness is constant, the length of a bed measured on this plane must therefore also remain constant.

Because of the requirement of the constancy of bed length, it follows that the length in cross section must be the same from one bed to another. It is then possible to apply a simple test of consistency to cross sections of parallel folds.

STEPS (after Dahlstrom, 1969a, p. 746)

1. Establish a pair of reference lines at either end of the section in regions of no interbed slip. These references lines should be located at the hinges of major anticlines or synclines, or in regions beyond the disturbed belt.

2. Measure the bed length of selected horizons between the reference lines. They should be the same.

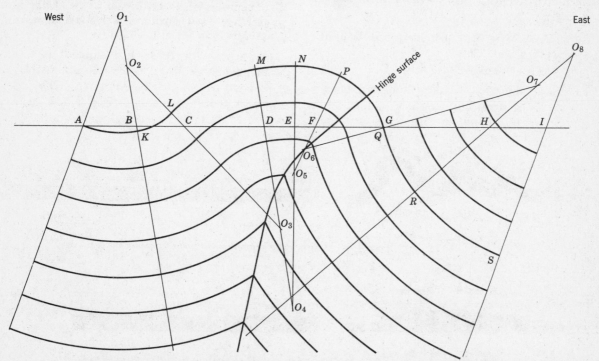

FIGURE 8.11 Full reconstruction of parallel folds, showing the trace of the hinge surface. (After Busk, 1929, p. 19.)

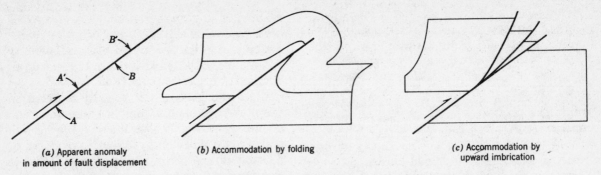

(a) Apparent anomaly in amount of fault displacement

(b) Accommodation by folding

(c) Accommodation by upward imbrication

FIGURE 8.12 Thrusts in parallel folds. (a) Apparent anomaly in amount of fault displacement. (b) Accommodation by folding. (c) Accommodation by upward imbrication. (From Dahlstrom, 1969a, Canadian Journal of Earth Science. Used by permission of National Research Council of Canada.)

3. If they are not the same, then the cross section must show a valid explanation of why they are not.

Sections which pass this test are termed *balanced* by Dahlstrom.

Parallel folds may be cut by thrust faults. If bed length remains constant, then the displacement on these faults should be consistent as well. However, the displacement commonly varies and the faults may die out altogether. There are only two ways of resolving this discrepancy (Dahlstrom, 1969a, p. 746):

1. Interchanging fold shortening and fault displacement (Fig. 8.12b).
2. Upward imbrication (Fig. 8.12c).

In these illustrations the fault traces are arbitrarily represented as straight lines. Such planar faults, together with constant bed length, require a substantial amount of interbed slippage which would alter the originally vertical ends of the blocks to the curved shapes shown. From these simple examples it may be concluded that (Dahlstrom, 1969a, p. 747):

1. Faults with changing displacement are apt to be curved, usually concave upward, a feature which is confirmed by observation.
2. Interbed slip is a necessary part of thrust faulting which accompanies parallel folding; and,
3. Interbed slip can contribute to the change in displacement, and in extreme cases would become a type of imbrication itself.

If the test for balance is applied to the parallel folds reconstructed by Busk's method of tangent arcs, it will be found that the bed length is not consistent (Fig. 8.13). The source of this discrepancy lies in the cusp-

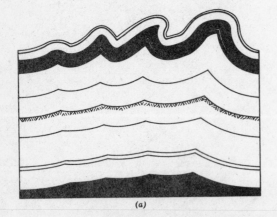

(a)

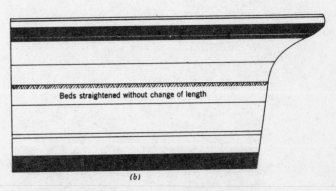

Beds straightened without change of length

(b)

FIGURE 8.13 Inequal shortening with depth. (a) Reconstruction by method of tangent arcs. (b) Beds straightened without change of length. (From Carey, 1962, Journal of the Alberta Society of Petroleum Geologists. Used by permission.)

shaped hinges in the cores of the anticlines. Clearly, these reconstructed forms can not represent the actual geometry of the deeper zones. This must mean that the parallel mechanism breaks down at depth, and that other types of deformation become important. The prediction of the geometry of these deeper structures from surface data in any detail is essentially impossible, and in many specific examples, is still being debated. However, the main alternatives can be outlined.

One reconstruction would be to keep the bed length constant with depth, but allow a small change in the thickness of each bed in order to keep the area constant (Fig. 8.14). The required changes in the form of the fold with depth is an example of *disharmony*, that is, folds that are out of harmony with the folded beds above.

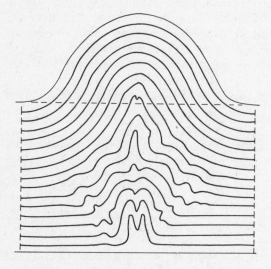

FIGURE 8.14 Reconstruction of folds at depth by maintaining original length and conserving area. (From Goguel, 1952, Traite de Tectonique, Masson et Cie. Used by permission.)

Another and much more fundamental structural change with depth is a direct consequence of this disharmony. How would the next lower bed of Fig. 8.14 be drawn? Clearly, the folded beds must have been completely detached from the underlying ones and deformed essentially independent of them. This requires a shearing-off horizon or *décollement* at the base of the figure.

The position of this basal detachment is determined by the location within the sedimentary sequence of weak layers, such as shale, or in extreme cases, salt and gypsum, or by the contact between the sedimentary rocks and an underlying rigid basement unit. In detail the behavior of the lower part of the folded sequence depends on the mechanical properties of the rocks involved. Instead of tightly crumpled folds in the cores of the anticlines, thrust faults may form that root in the décollement zone. These faults may or may not break through to the surface, and if they form early in the folding they themselves may be deformed by continued folding or by later thrusting.

Finally, in terms of cause and effect, it is the position of the detachment plane, in combination with the actual shortening, that determines the size and shape of the folds in the fold belt. The thicker the strata above the potential detachment the larger the individual fold size. A comparison of two folds in the Jura Mountains illustrates this control (Fig. 8.15).

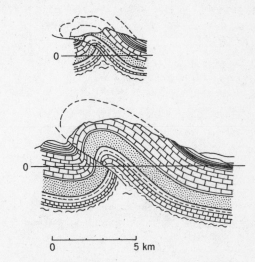

0   5 km

FIGURE 8.15 Folds from the Jura Mountains with size proportional to thickness. (From De Sitter, 1964, Structural Geology, McGraw-Hill. Used by permission.)

The effect of folding is to make a packet of rocks thicker and shorter. If volume is conserved in the packet, the amount of material uplifted must exactly equal the decrease due to shortening. Since both the amount of thickening and shortening can be measured on a cross section, the depth to the décollement can be determined.

CALCULATION (after De Sitter, 1964, p. 190; Dahlstrom, 1969b, p. 342)

1. The lateral shortening of a reference horizon between points of no slippage is determined by comparing the bed length measured with a curvimeter, with the lateral distance this horizon occupies in the folded packet. In Fig. 8.16b, the bed length measured at the top of the Mississippian Rundle group is $AEFG = AB$. The distance GB represents the actual shortening.

2. The amount of increased area of the thickened packet between the two no-slip reference points is measured with a planimeter. In Fig. 8.16b, this area lies between the straight line AG and the trace of the folded reference horizon $AEFG$.

3. The depth of folding follows directly from the relationship:

Depth AD = (area uplifted) / (shortening)

Note that the area of rectangle $GBCH$ equals the area measured in step 2.

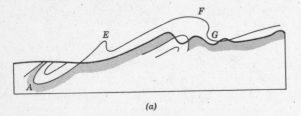

(a)

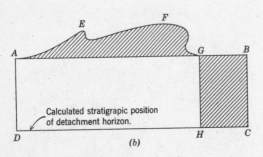

Calculated stratigrapic position of detachment horizon.

(b)

FIGURE 8.16 Calculation of depth of folding. (After Dahlstrom, 1969b, Bulletin of Canadian Petroleum Geology. Used by permission.)

The method has been used with good results in the Jura Mountains where the concept of the décollement originated (see Bucher, 1933, p. 151f). However, these results are not without difficulties. How is it that a packet of

sedimentary rocks can be deformed independently from the underlying material? Suggestions have been made concerning the possible existence of related structures, such as imbricate slices in the underlying block. Besides raising several additional difficulties, there is increasing evidence that there is in general no sub-décollement involvement in many fold belts, but the agent responsible for the deformation remains in question (see Pierce, 1966).

It should be abundantly clear that parallel folding is a complicated process and must involve other modes of deformation, including disharmony and shearing-off. It does not follow, however, that every group of parallel folds has a single décollement thrust at depth. The necessary adjustments may take place locally and gradually rather than at a single horizon. More importantly, the rocks that underlie the parallel folds may indeed be deformed. Examples of flow during conditions of high temperature and pressure are present in every metamorphic terrane. Synchronous folding of overlying, near surface layered rocks by essentially parallel processes would be a certainty, and in general there would be a vertical transition between the two realms.

Barnes and Houston (1969) have described a simple example which illustrates the principal involved. In a part of the Medicine Bow uplift in the Northern Rocky Mountains of Wyoming, a Precambrian metamorphic basement is unconformably overlain by Paleozoic and Mesozoic sedimentary rocks. During the Laramide these layers were folded (Fig. 8.17), presumably in response to distributive movement on microfractures in the basement unit. Under these circumstances shearing-off is not required. Compton (1967) described a similar example where he was able to demonstrate the actual slip on closely spaced fractures in a gneissic basement. Up to 3700 m of overlying sedimentary rocks were deformed by folding. An interesting feature is that there is evidence of disharmony in the *upper* part of the sequence, particularly in the cores of synclines. Dahlstrom (1969b) has shown that in certain circumstances this upward increase in disharmony may actually result in an upper detachment fault.

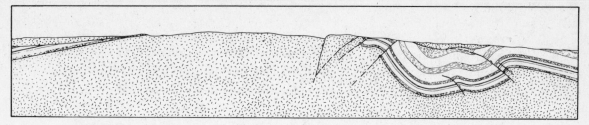

FIGURE 8.17 Laramide folds in the Northern Rocky Mountains. (From Barnes and Houston, 1969, Contributions to Geology. Used by permission.)

NONPARALLEL FOLDS

The breakdown of the ideal parallel fold geometry with depth implied in the disharmony-décollement relationships is, in any detailed form, unpredictable, simply because it follows no geometrical rules that can be deduced from surface observations. In addition, however, nonparallel modifications of folds in sedimentary rocks may be observable in the field.

There is a limit to the amount of shortening possible in a parallel fold. Theoretically, bedding plane slip must cease when the two limbs have been rotated into parallelism. In Fig. 8.18, the length of the semicircular arc is πr, and occupies a horizontal distance of $2r$. From the definition of extension, $e = (2r - \pi r)/\pi r = 0.36$, or a maximum of 36% shortening. If the folds are further deformed, certain beds will begin to thin, and the fold will depart from strict parallelism. The closeness to the theoretical limit at which this thinning occurs depends on the mechanical properties of the rock layers undergoing folding. After a certain degree of shortening, the parallel fold usually becomes asymmetrical, and one limb becomes steeper than the other, and will finally become overturned. Although it may occur earlier, at the point of overturning thinning of the steep limb is geometrically required (Busk, 1929, p. 30). The less steep limb may still be parallel. In terms of reconstruction by circular arcs, this nonparallelism is proved when correlation of a key horizon and the utilization of certain dip measurements are irreconcilable. The simplest approach in such a case is to make the necessary adjustments in the thinner limb by freehand sketching (Fig. 8.19).

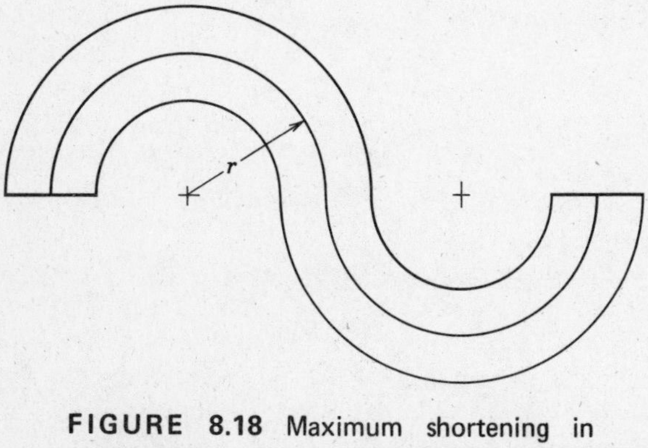

FIGURE 8.18 Maximum shortening in parallel folds.

FIGURE 8.19 Sketch of parallel fold modified by thinning of overturned limb. (After Busk, 1929, p. 57.)

EXERCISES

1. On a card deck inscribe an array of small circles, and with this deck produce a flextural slip fold. Confirm that it has parallel form, and that the relationship between strain and limb inclination illustrated in Figs. 8.3 and 8.4 holds.

2. The traverse data given below was obtained along a west to east traverse. Reconstruct the folds, calculate the depth of folding, and sketch the possible structure in the deeper parts.

Station	Distance	Elevation	Dip
A	0	650 m	30 E
B	900	800	41 W
C	2300	850	18 W
D	2550	750	37 E
E	3350	800	44 W
F	4900	1150	5 E
G	5350	1000	69 E
H	6900	650	8 W
I	8000	550	66 W

9
Similar Folds

Several explanations of similar folds have been proposed (Ramsay, 1967, p. 43; Bayly, 1971; Matthews, et al., 1971). The simplest of these can be illustrated with a second card deck analog. If a layer is represented by a band drawn on the edge of a card deck, and the deck is deformed by inhomogeneous simple shear, the band will be folded. This process is called shear folding. Since each card and the portion of the band marked on its edge remains undistorted during the displacement, the thickness of the band measured parallel to the shear plane is constant. Also because this dimension remains unchanged, the slope of the band boundary defined on two adjacent cards is the same on both the inner and outer curves of the model fold. This means that two bounding curves are identical and that both the dip isogons and the traces of the hinge and inflection surfaces are parallel to the edge of the cards. The property of parallel isogons and of constant axial plane thickness are distinguishing characteristics of similar folds.

In describing the geometry of shear folds it is convenient to establish a coordinate system which is directly related to the geometry of simple shear. The shear plane is usually taken as the ab-coordinate plane, with the a-axis being parallel to the direction of shear. In the card deck models, the face of the deck on which the drawings are made is the ac-plane,

and the edge of the cards on this plane is the a-direction.

If a layer is originally parallel to the bc-plane, and if the form of the displacement is

$$a = A \sin c \qquad (9.1)$$

where A is the amplitude, then both bounding curves will have this same form (Fig. 9.1). However, other original attitudes and displacement curves are possible. For example,

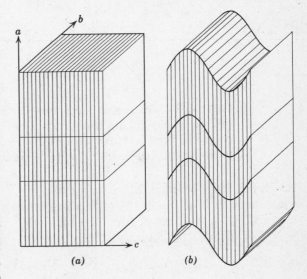

FIGURE 9.1 Card deck model of a shear fold.

71

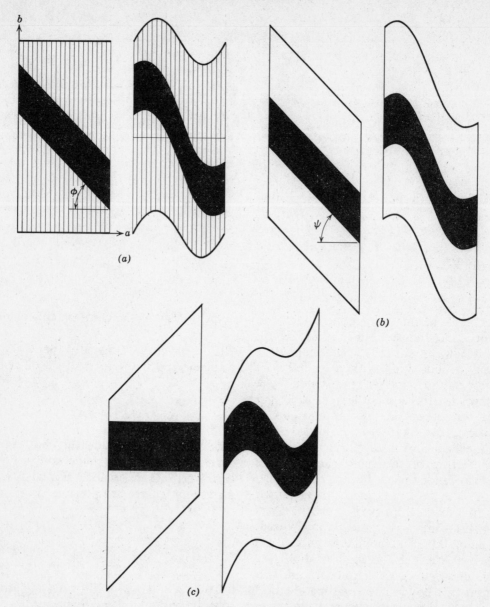

FIGURE 9.2 Symmetric and asymmetric similar folds. (*a*) Symmetric displacement of an originally oblique layer produces an asymmetric fold. (*b*) Addition of a component of homogeneous simple shear destroys symmetry of fold. (*c*) A special combination of homogeneous and inhomogeneous shear produces a symmetric fold.

the layer may be inclined to the *c*-axis at some angle ϕ (Fig. 9.2*a*). If the displacement curve is still $a = A \sin c$, then the fold will have the form (after Ramsay, 1967, p. 426)

$$a = A \sin c + c \tan \phi \qquad (9.2)$$

This case brings out two additional features: (1) the axial plane thickness is generally not

the thickness of the original layer; and, (2) the hinge of the fold may be shifted laterally from the maximum and minimum on the displacement curve.

This same result may be obtained somewhat differently. A band originally parallel to the *c*-direction (as in Fig. 9.1) may be rotated into an inclined orientation by homogeneous simple shear, and then distorted by the $a = A \sin c$ type of inhomogeneous strain (Fig. 9.2b). The total displacement curve, and the resulting fold have the form

$$a = A \sin c + c \tan \psi \qquad (9.3)$$

where ψ is the angle of shear of the homogeneous component. The order of the stages of deformation may be reversed to achieve this same result, or the two components may act together during a more complex pattern of displacement. Finally, an original inclination (as in Fig. 9.2a) may be eliminated by homogeneous simple shear, and the $a = A \sin c$ displacement then superimposed to produce a fold (Fig. 9.2c). The total displacement curve is the same as (9.3), but because the original slope of the band and the slope of the line defining the component of homogeneous simple shear are equal and opposite, the resulting fold has the simpler symmetrical form of (9.1).

From a comparison of the three examples illustrated in Fig. 9.2, and the accompanying equations, it should be apparent that, in terms of the final fold form, the effect of a component of homogeneous simple shear can not be separated from the effect of original inclination, and that a considerable amount of strain may be effectively concealed. It should also be clear that there are an infinite number of ways of accommodating the variations in the states of strain within the layer. This is true for all fold types, and points up an important limitation of the classification based on geometry of the folded layers (Ramsay, 1967, p. 344).

Two objections have been raised concerning this shear folding mechanism. First, folds often termed *similar*, but which do not possess the ideal geometry have been shown to result from an entirely different mechanism—the homogeneous flattening of originally parallel folds (Ramsay, 1967, p. 441). This difficulty is avoided by reserving the term similar for the special class of fold shape, rather than applying it to all folds with pronounced thinning of limbs, a characteristic also present in folds with weakly convergent and divergent isogonal patterns.

The second objection, and a more serious one, is directed at the mechanical feasibility of the systematic reversal in the sense of shear as required in the examples given here. However, under certain circumstances, single-sense inhomogeneous simple shear is also capable of producing well-developed similar folds (Fig. 9.3). For a fold of the type $a = A \sin c$, the slope at the limb inflection point a_i is the maximum, for example, if $A = 1$, $a_i = 45°$, if $A = 2$, $a_i = 60°$, etc. For single-sense shear to produce such a fold, the initial angle of inclination of the band $\phi \geqslant a_i$, and $\psi \geqslant \phi$, but of opposite slope. This mechanism seems to operate in glacier ice (Ragan, 1969b).

FIGURE 9.3 Single sense shear to produce a well developed fold.

In three dimensions, the relationship between similar folds and shear geometry is even more involved. In similar folds the axial plane coincides with the shear plane (= ab-plane), and the hinge line and the fold axis are always parallel to the intersection of the shear plane and the surface being folded. Therefore, if the original layer is inclined to the b-axis at some angle β, the fold axis will also be inclined at β, and the orientation of the fold axis can *not* be used to determine the a direction (Fig. 9.4). Since the profile plane of the fold is no longer parallel to the ac-plane, the fold shape will always be a subdued version of the displacement curve.

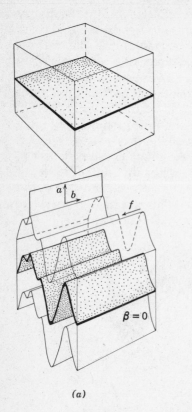
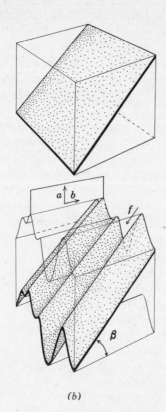
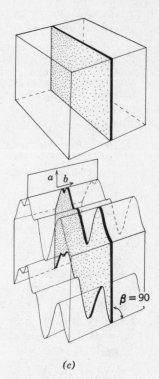

(a)   (b)   (c)

FIGURE 9.4 Similar folds developed in layers with variable inclination to the *b*-axis. (a) Original layer parallel to the *b*-axis. (b) Layer inclined to the *b*-axis at angle β (c) Original layer perpendicular to the *b*-axis. (From Ramsay, 1967, p. 425-426. Fracturing and Folding in Rocks. McGraw-Hill. Used by permission.)

The amplitude A of the fold measured in the profile plane is related to the amplitude A_{max} of the displacement curve by

$$A = A_{max} \cos \beta \qquad (9.4)$$

The limiting case occurs when $\beta = 90°$, and $A = 0$; that is, when the layer is parallel to a no fold develops at all.

SUPERPOSED FOLDS IN TWO DIMENSIONS

Card deck models can also be successfully used to illustrate the effect of shear folding on previously existing folds. The form of the first folds is, as before, simply drawn on the edge of the pack, the deck deformed inhomogeneously (Fig. 9.5). During this deformation

the originally straight traces of the first hinge surfaces (HS 1) behave in the same fashion as

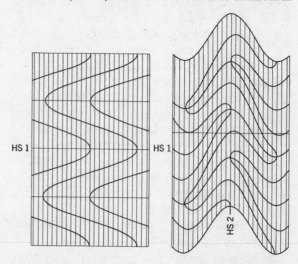

FIGURE 9.5 Model of superposed similar folds.

the layer boundaries of the previous examples. At the same time, the first folds are complexly recurved. Two features are noteworthy: (1) the hinge points of the first folds do not coincide with the points of maximum curvature on the second folds, and (2) the trace of the hinge surface of the second displacement curve (HS 2) does not pass through the points of maximum curvature on the limbs of the now twice folded bands, but alternates from one side to the other as it passes from limb to limb.

Given an example of such superposed folds, it is possible to unravel the two stages of deformation using the simple rules of shear folding (see Fig. 9.6).

FIGURE 9.6 Example problem of the analysis of superposed folds.

ANALYSIS (after Carey, 1962)

1. *Identify the first generation folds*: On a tracing of the superposed folds number the layering in sequence (the relative ages may not and need not be known). If continuity is lost in highly attenuated zones it is usually possible to work around them; if not, a second or even third sequence may be started using Roman or Greek letters. Once all units are labeled, special patterns will identify the cores of the first folds. For example, at point *X* (Fig. 9.7) the numbers run 3 4 5 4 3, and at point *Y*, 9 8 7 8 9. Although without relative ages they can not be

distinguished, clearly one of these cores represents an anticline and the other a syncline.

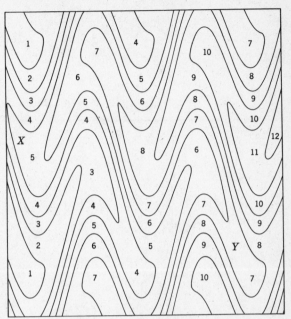

FIGURE 9.7 Tracing and numbering.

2. *Involuted hinge surfaces of the first folds*: Wherever the core layers form an apex, the trace of the folded hinge surface of the first fold (HS 1) must pass into the next layer. Since the first hinge points can not be accurately located, the point of maximum curvature is used as a close approximation. Through these approximate hinge points draw in the traces of the first hinge surfaces using solid, dashed or dotted lines to indicate the degree of confidence in their location. These traces can then be marked with zeros or cross-bars depending on whether the index numbers are minimum or maximum in the fold cores (Fig. 9.8).

3. *Hinge surfaces of the second folds:* A second set of traces can be drawn through the hinge points of the folded limbs. These appear as a series of roughly straight and parallel lines. Using crosses and zeros to mark sequences which rise or fall it will be seen that groups of symbols alternate along these traces. If difficulty was encountered in drawing the involuted HS 1 traces through strongly attenuated zones or across widely spaced areas, they may be completed using the following clues:

 a. Involuted HS 1 traces of like sign must join.
 b. The sense of curvature of the HS 1 traces must be the same where they cross a particular HS 2 trace, that is, all must be either concave up or down.

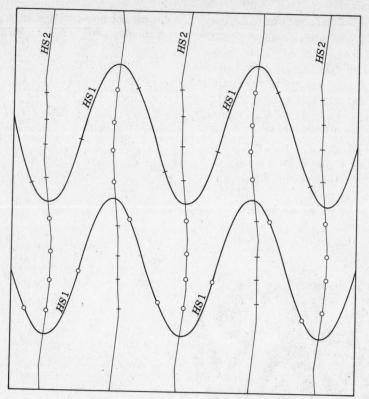

FIGURE 9.8　Traces of *HS* 1 and *HS* 2.

c. The folded HS 1 traces cross the straight HS 2 traces only at points where the sign changes occur. This fact may also be used to determine the number of HS 1 traces to be inserted across gaps.

4. *The directrix and form of the second folding*: By collecting all the HS 1 and HS 2 lines on a single tracing, and by visually averaging both, the displacement curve and the directrix responsible for the second folding can be extracted from the complex pattern (Fig. 9.9).

5. *Form of the first folds*: The hinge surfaces of the first folds are assumed to have been planar. Therefore the effect of the second folding can be eliminated by shifting the structure parallel to the displacement curve derived from the original superposed folds (Fig. 9.10). This shifting is most easily accomplished by:

a. Draw a series of closely spaced lines on the original fold pattern parallel to the directrix of the second folding.

b. On a tracing sheet, draw a second set of parallel lines with the same spacing.

c. Overlay this sheet on the fold pattern and mark off the points of intersection of the folded bands with the first line, and then shift the tracing according to the reverse displace-

ment curve and repeat for the second line, and so forth.

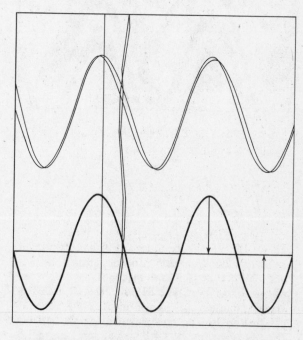

FIGURE 9.9　Determination of *F*2 fold shape.

The pattern of the first folds can be drawn by connecting points across the guidelines. The first attempt is apt to be somewhat crude. Irregularities due to drafting and positioning errors may be filtered out by repeating the tracing process illustrated in Fig. 9.9. The smoothed version will appear as in Fig. 9.10.

(a)

FIGURE 9.11 Multiple superposed folds. (a) three sets, (b) four sets.

FIGURE 9.10 Unfolding by straightening the *HS* 2 traces.

WILD FOLDS

In theory and in concept, this process of superimposing folds upon folds could be repeated any number of times. To model such multiple folds with the aid of card decks would require that the convoluted patterns of earlier experiments be transferred onto a deck with a different orientation relative to the *a*-direction. Unfortunately, any attempt to perform such an experiment meets with severe practical difficulties which mount exponentially beyond the second set. However, such patterns are easily produced with

(b)

the aid of a computer-driven plotter. The technique consists simply of adding sine curves with varying amplitudes and wave lengths alternately along the x- and y-axes. Fig. 9.11 shows typical runs for three and four sets of superposed folds.

Irregular folds are characterized by irregularity of axial planes and discontinuities and rapid variations in thickness of bands (Fleuty, 1964, p. 477). The most disordered types also show a wide variation in attitudes of hinges and hinges surfaces. Such folds are particularly common in migmatic gneisses, where they give the appearance of stirred porridge (De Sitter and Zwart, 1960, p. 253), and are sometimes called *wild* folds (Kranck, 1953, p. 59; Berthelsen, et al., 1962). Except for their periodic character and perfect continuity, certain aspects of these computer generated patterns are similar to the wild folds found in nature, suggesting that their appearance may be a matter of complexity rather than irregularity.

SUPERPOSED FOLDS IN THREE DIMENSIONS

Card-deck models can also be used to illustrate superposed folds in three dimensions. The first folds are represented by a cylindrical surface cut across the deck; deformation of the deck by inhomogeneous simple shear then refolds this structure. Although the method requires special preparations, including the cutting and deforming of the cards, a number of interesting and informative experiments can be performed which are well worth pursuing further (see O'Driscoll, 1964). A simple example will indicate the approach and its potential. If original upright folds are deformed by a second set of upright folds trending at right angle to the first, a series of domes and basins result (see Fig. 9.12). Other angles between the first and second folds can be simulated by homogeneously shearing the deck in a horizontal direction; the domes and basins are then asymmetric and en échelon.

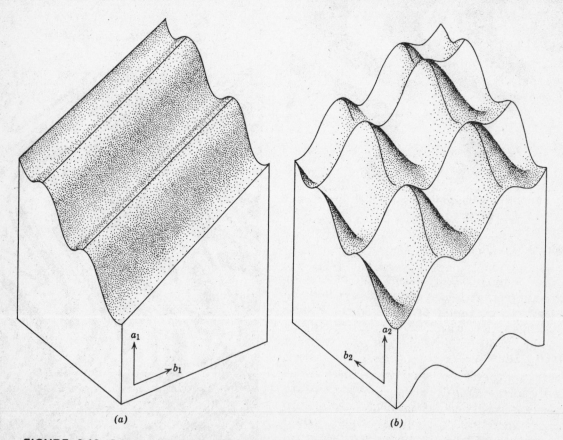

(a)   *(b)*

FIGURE 9.12 Superposed folds in three-dimensions. (After O'Driscoll, 1962, p. 166.)

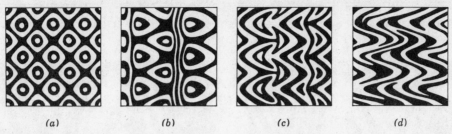

FIGURE 9.13 Interference patterns of superposed folds. (From Ramsay, 1967, p. 531. Fracturing and Folding in Rocks. McGraw-Hill. Used by permission.)

Because of their likeness to the patterns caused by the intersection of two sets of waves, these are called *interference patterns*. If instead of a single folded surface, the deformation operated on a multilayered block, and if an exposure plane cuts horizontally through the superposed folds, an outcrop pattern of the interference structures results. The character of this type of pattern depends on the relative orientation and size of the first and second folds. If, as in the model, the second folds are horizontal and upright, and further, if the sizes of both sets of folds are assumed to be the same, the resulting pattern can be described directly in terms of the attitude of the first folds. Within this framework, several types of patterns can be distinguished.

a. If the first folds are horizontal and upright, and the trends of the two folds are perpendicular, the patterns of domes and basins alternate with a high degree of symmetry and regularity (Fig. 9.13a).

b. Inclined first folds are reflected by dome-basin patterns which no longer have simple rounded forms. This reflects the original difference in dip on the limbs of the first folds (Fig. 9.13b).

c. Inclined first folds with an overturned limb result in characteristic crescent and mushroom patterns (Fig. 9.13c).

d. If the first folds are recumbent or reclined, and the trends of the two sets are parallel, the patterns do not differ in kind from the two dimensional patterns which are so easily modeled with card decks (Fig. 9.13d).

Other relative orientations give patterns which are transitional with these four, and many can be illustrated by means of card-deck experiments as described by O'Driscoll.

EXERCISES

1. With a card deck produce the following types of shear folds. Across one end of the deck add a row of small circles, and observe the strain distribution in each experiment.
    a. A symmetrical sinusoidal fold with reverse sense of shearing (as in Fig. 9.1).
    b. Add a component of homogeneous simple shear to produce an asymmetric fold (Fig. 9.2).
    c. A symmetric fold by single sense shearing (Fig. 9.3).

2. In the light of these experiments, write a short statement concerning the indeterminant relationship between final fold geometry and the strain distribution.

3. Draw a similar fold on the card deck with the trace of its hinge oblique to the direction of shear. Deform the deck to produce a superposed fold, and note the thickness variation in the now twice folded layer, and the location of new hinge points on the old fold limbs and the locations of the hinge points of the first folds.

4. Using Fig. X9.1 (see p. X-5) remove the effects of F2 folding to give the form of the F1 folds. As will soon become apparent, the F1 folds in this problem are

highly regular, and it is only necessary to proceed to the point where their form can be confirmed with some confidence, rather than attempting a complete re-construction.

FIGURE X9.1

10
Folds
and Topography

MAP SYMBOLS FOR FOLDS

If exposures are good and the folds small enough, the shape of the fold, and thus the location of the hinge surface, may be determined directly in the field. If not, they must be reconstructed from attitude data and the mapped trace of the folded layers. In either case, the fold is depicted on the map with a line marking the trace of the hinge surface together with its attitude. Similarly, the attitude of the hinge line is determined, directly or indirectly, and its trend and plunge plotted on the map. If the hinge surface is vertical, the trend of the hinge line is parallel to the strike of the hinge surface, although it may plunge at any angle. In the general case of an inclined fold, there is always an angle between the trend of the hinge line and the strike of the hinge surface. Map symbols covering all these situations are shown in Fig. 10.1.

OUTCROP PATTERNS

Just as structural planes intersect the earth's surface to give characteristic outcrop patterns,

equally distinctive though more complex patterns result from the intersection of fold with the topographic surface. Horizontal folds are a commonly occurring and a simple type of fold, and they also have the simplest type of outcrop pattern. On a horizontal exposure plane, the pattern of such folds is essentially the sum of the patterns of the inclined limbs, that is, a series of parallel outcrop bands (Fig. 10.2). Without knowledge of the relative age and correlation of the units, or the attitude at several points, it may be difficult or impossible to interpret this pattern as being part of a folded structure at all. However, once these are known, the existence of a fold becomes clear. If the same fold is exposed in an area of some topographic relief, the dips on the opposing limbs are immediately evident, and the interpretation of a fold can be made with some confidence. If the topography is such that the hinge intersects the surface, the fold becomes obvious (Fig. 10.3).

For plunging folds, a converging pattern is characteristic and unmistakable (Fig. 10.4). However, on the basis of pattern alone it is impossible to distinguish antiforms from synforms. Again, if the dip at several points, the direction of plunge, or the sequence is

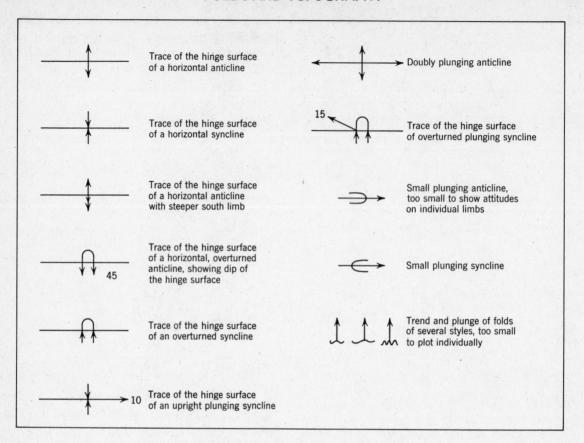

FIGURE 10.1 Map symbols for folds.

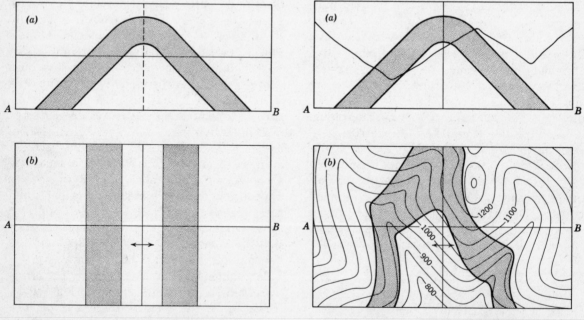

FIGURE 10.2 Horizontal upright fold: (a) in profile, (b) exposed on a horizontal plane surface.

FIGURE 10.3 Horizontal upright fold (same as Fig. 10.2): (a) in profile, (b) exposed in area of topographic relief.

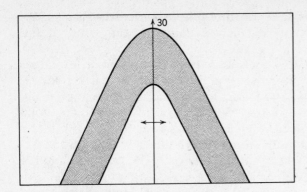

FIGURE 10.4 Plunging upright fold exposed on a horizontal plane surface.

known, the type of fold may be immediately determined. Some of this information is supplied if the plunging fold is exposed in an area of topographic relief (Fig. 10.5).

In all these cases, given attitude data, vertical cross sections could be constructed. Only if the fold is horizontal does such a section give authentic information concerning fold geometry. Plunging folds are always distorted in vertical sections: the shape of the folded surfaces, the thickness variations in the folded layers, and the interlimb angles are all

incorrectly shown. If the aim is to portray fold geometry, such vertical sections are worse than useless, and must be avoided. It is necessary to construct the plane perpendicular to the fold axis, that is, the *profile* plane.

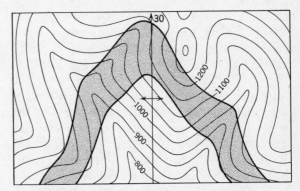

FIGURE 10.5 Plunging upright fold exposed in area of topographic relief.

DOWN-PLUNGE VIEW OF FOLDS

By viewing map patterns of tilted, but not folded, strata in the direction of dip an

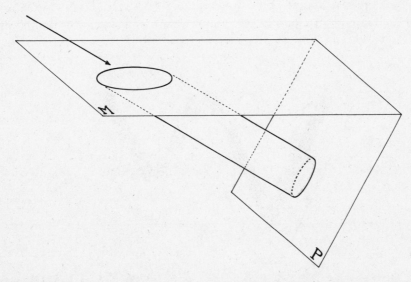

FIGURE 10.6 The down-plunge view of a circular cylinder: *M* = map plane, *P* = profile plane.

important visual simplification is achieved (Chapter 3). This method may be used to even greater advantage in viewing plunging folds. By turning the map and adopting a view so that the line of sight is in the direction of the plunging fold axis, the profile of the fold is actually seen. As before, the principle is that the distortions at the earth's surface are eliminated by this down-plunge view. This is easily seen in the case of an inclined cylindrical pipe of circular cross section (Fig. 10.6). If the pipe is truncated at a horizontal surface its form will be an ellipse. If this ellipse is then viewed parallel to the inclined axis of the pipe, it will again appear to be circular. Fig. 10.7 shows a map of several upright plunging folds. With a down-plunge view, the antiforms and synforms are simply seen as such. The map pattern also indicates that disharmonic folds are also present, and the down-plunge view automatically includes them in their proper place in the structure. In contrast, a section mechanically constructed along line SS' would fail to represent the small plications in the cores of the folds.

If the folds are inclined, the correct shape of the plunging folds is seen only if the line of sight is parallel to the fold axis, not when it is parallel to the trace of the hinge surface. Fig. 10.8 illustrates this rule. As shown on the map (Fig. 10.8a), the folds plunge due north at an angle of 20°. Though not shown, the trace of the hinge surface trends N 15 E, and dips steeply to the west. With a line of sight parallel to the fold axis, the true shape of the folded layers is revealed; the result of this view is also shown on the profile (Fig. 10.8b). The folds are clearly of parallel type, and are overturned to the east. Note that if folds are viewed parallel to the strike of the hinge surfaces, that is, toward N 15 E, the folds falsely appear upright, with notable thinning on the steep eastern limbs of the antiforms.

The down-plunge view reveals another important feature of fold patterns. The folds of Fig. 10.7 have vertical hinge surfaces, and thus the trace of this surface also connects points of greatest curvature of the outcrop pattern, a fact that can be readily seen in the down-plunge view. This coincidence holds *only* when they are exposed in areas of negligible topographic relief. If the relief is significant there will, in general, be a discrepancy between the real and apparent hinge points (see Fig. 10.3b, 10.5). For inclined plunging folds exposed on horizontal plane surfaces, the line connecting the points of greatest curvature of the outcrop pattern and the trace of the hinge surface always depart. The amount of the departure depends on the attitude of the fold and the nature of the curved boundary surfaces, and the effect is compounded by topographic relief. For similar folds the two lines are parallel but not coincident (Fig. 10.9a), and for parallel folds the lines are neither coincident nor parallel (Fig. 10.9b). If this relationship seems difficult to accept, a down-plunge view of the

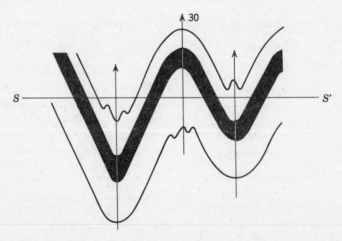

FIGURE 10.7 Map pattern of upright plunging folds. (From Mackin, 1950, Journal of Geology. Used by permission.)

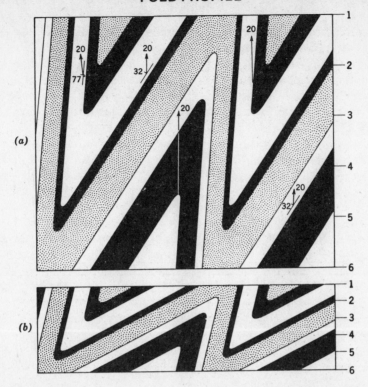

FIGURE 10.8 Map pattern of gently plunging, steeply inclined folds and profile drawn perpendicular to fold axis. (From Mackin, 1950, Journal of Geology. Used by permission.)

folds of Figure 10.9 will immediately confirm its validity.

FOLD PROFILE

The true form of cylindrical folds seen in the down-plunge view may also be constructed from a geologic map. There are two different, though equivalent, construction techniques. The first is the more straightforward one, and applies to areas of negligible relief. It involves constructing the foreshortened map pattern with the aid of a grid (after Wegmann, 1929).

CONSTRUCTION (Fig. 10.10)
1. On the geologic map, draw a square grid with one coordinate direction parallel to the trend of the fold axis (Fig. 10.10a).
2. Viewed down-plunge, the grid spacing across the line of sight, that is, perpendicular to the fold axis, remains unchanged (1, 2, 3, . . .). The other, in the direction of the axial trend, is reduced to $k \sin p$, where k is the original grid spacing, and p is

the angle of plunge. This new grid spacing may be computed, or found graphically (Fig. 10.10b).
3. A second grid is then constructed representing the down-plunge view: the spacing 1, 2, 3, . . . is the same as originally constructed on the map, while a', b', c', . . . is the foreshortened one.
4. The fold pattern is then manually transferred from the map grid to the profile grid (Fig. 10.10c).

The resulting fold profile involves no speculation; no line appears on it that is not also on the map. Contrast this with the construction of a vertical cross section of parallel folds in which surface attitudes are projected at right angle to the fold axis. Both profiles and vertical sections require projection of data. For the profile, however, the data are projected parallel to the fold axis, a direction of minimum change in cylindrical folds, whereas the vertical section requires projection in the direction of maximum change.

The second technique, using an orthographic construction, is more involved, but is adaptable to complex situations (after Wilson, 1967).

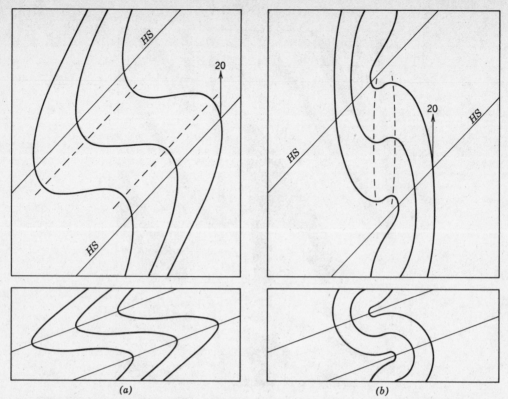

FIGURE 10.9 Map and profile view demonstrating the lack of correspondence between the trace of the hinge surface *HS* and the line connecting the points of greatest curvature of the outcrop pattern. Both sets of folds plunge 20° due north (*a*) similar folds, (*b*) parallel folds. (After Schryver, 1966.)

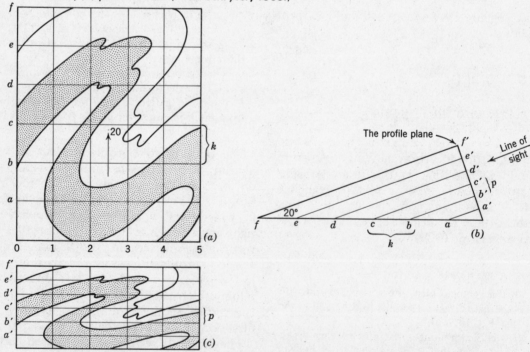

FIGURE 10.10 Construction of fold profile. (*a*) Geologic map with grid. (*b*) Calculation of the foreshortened grid. (*c*) Map pattern transferred to the new grid to give the down-structure view.

CONSTRUCTION

1. Establish folding lines FL 1 parallel and FL 2 perpendicular to the trend of the fold axes, preferably outside the map area (Fig. 10.11a).

2. Folding about FL 1, draw a vertical section parallel to the fold axes. In this section the line OP represents the profile plane in edge view. A series of selected points on the map are projected to FL 1, thence to OP using the angle of plunge.

3. Folding about FL 2, these points are projected from both the map and from OP to fix their location on the profile. See the details of the projection of point 1 on the map to point 1' on the profile (Fig. 10.11a, b).

4. After a sufficient number of points have been transferred in this manner, the form of the folds on the profile can be completed by connecting appropriate points.

5. The result is an "up-plunge" view, but this can be easily converted to a down-plunge one simply by reversing it (Fig. 10.11c).

The advantage of this method is that fold profiles can be also constructed when the map area has topographic relief.

CONSTRUCTION

1. As before, construct folding lines FL 1 parallel and FL 2 perpendicular to the trend of the plunging folds.

2. Instead of projecting directly to FL1, the topography must be taken into account by adding a series of elevation lines representing the contour interval of the topographic map, plotted using the map scale (Fig. 10.12a). The points are projected to their corresponding elevation lines.

3. These points are then projected to the profile plane OP, using the angle of plunge. The fold profile is drawn as before from points projected from this edge view and from the map. This final projection can be made easier if the use of circular arcs is avoided. By constructing line OB to bisect angle POR, the points can be projected along the plunge lines directly to this bisector, and then to the profile, saving one step (Fig. 10.12b).

Strictly, the construction of a fold profile requires that the fold axes of the entire area be constant in trend and plunge, or at least have negligible variability. If the axes are not constant, then the folds are not cylindrical, and no single direction of view or projection exists. However, if the plunge angle changes progressively over an area, it is possible to draw an approximate profile, and thus to dipict the general fold style. One method is to draw a series of overlapping strip profiles for small areas where the plunge is essentially constant, and join them to make a composite

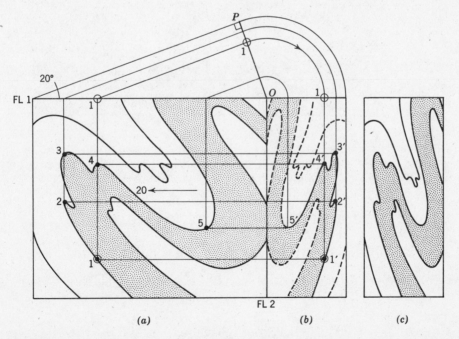

(a) *(b)* *(c)*

FIGURE 10.11 Construction of the fold profile by orthographic methods. (a) Map with folding lines parallel and perpendicular to the fold trend. (b) An "up-structure" profile. (c) The down-structure profile by reversal.

section. A second method is to adapt the orthographic construction used above.

CONSTRUCTION

1. Proceeding as before, establish folding lines FL 1 and FL 2 (Fig. 10.13a).
2. Project the data points to FL 1 and for each,

plot the associated plunge angle. With the tangent arc method used in reconstructing parallel folds (Chapter 8), curved plunge lines are drawn to project the points to the profile plane.

3. From the map and the edge view OP, points are there projected to the profile plane, and the form of the folds completed by sketching between control points.

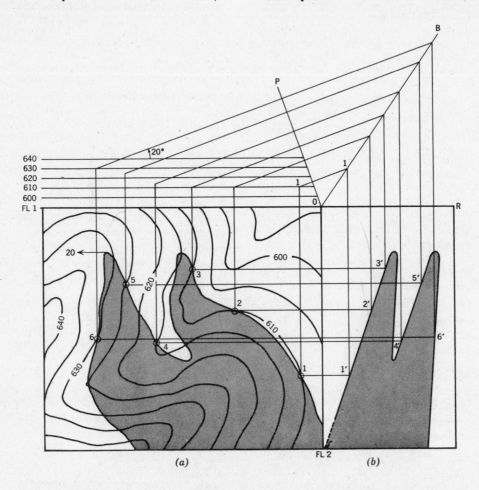

FIGURE 10.12 Fold profile constructed from outcrop pattern in area of topographic relief: (a) geologic map, (b) fold profile.

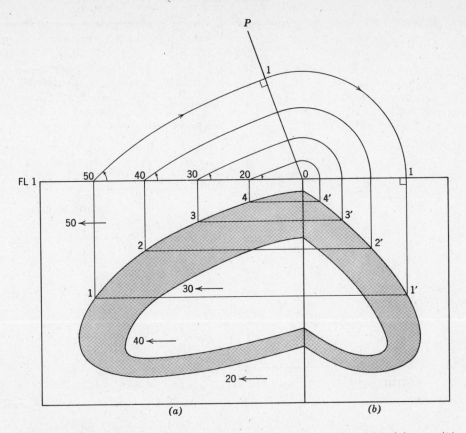

FIGURE 10.13 Approximate fold profile of a variably plunging fold: (a) map, (b) profile.

EXERCISES

1. The folds depicted in Fig. X10.1 (see p. X-6) plunge 30° due east (the method of determining fold attitude will be treated in Chapter 13). Add the appropriate map symbols (see Fig. 10.1).

2. The fold in Fig. X10.2 (see p. X-7) plunges 25° due north. Construct a profile of the fold perpendicular to the plunge and add the appropriate symbols to the map.

3. Using the geologic map of the Coatesville-West Chester District, Pennsylvania (Fig. X10.3, see p. X-8) assume the attitude of the fold axes is 15, S 65 W, and draw a profile of the structure. Concentrate on the main elements of the pattern, rather than on the details. Noting the order of superposition, describe the structure. In the light of these results do you think it necessary to interpret any part of the map pattern as being due to superposed folding? It is of some interest to pursue two opposing views on this question: see McKinstry (1961), and Mackin (1962).

FIGURE X10.1

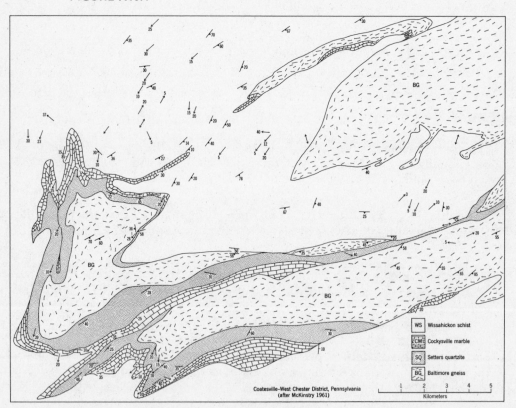

FIGURE X10.2

FIGURE X10.3

Coatesville–West Chester District, Pennsylvania
(after McKinstry 1961)

WS	Wissahickon schist
CM	Cockysville marble
SQ	Setters quartzite
BG	Baltimore gneiss

Kilometers 1 2 3 4 5

11
Graphic Solutions with the Stereonet

STEREOGRAPHIC PROJECTION

The solution of problems by the usual methods of descriptive geometry requires the construction of at least two orthographic views, and this consumes much time and effort. Fortunately, there is an alternative approach by which the *angular* relationships of lines and planes can be determined more readily.

If a sphere is constructed centered at some point O on the outcrop trace of an inclined structural plane (Fig. 11.1a), the plane and its extension will intersect this sphere as a great circle (Fig. 11.1b); that is, one whose plane includes the center point O. In order to be of

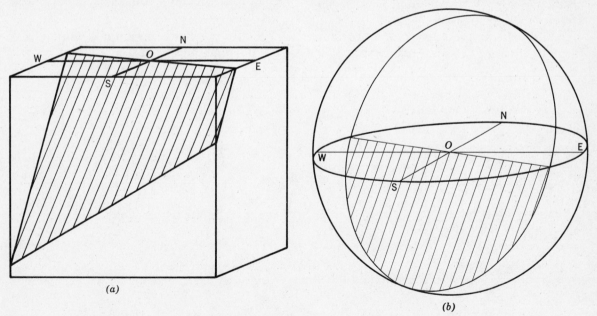

(a)

(b)

FIGURE 11.1 Representation of an inclined plane. (*a*) Block diagram with point O on the outcrop trace. (*b*) A sphere constructed with center at point O. (**After Phillips, 1971.**)

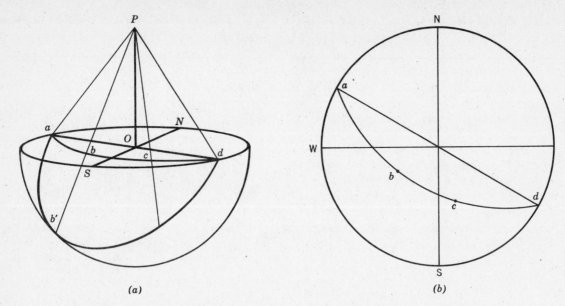

(a)

(b)

FIGURE 11.2 Stereographic projection of an inclined plane. (a) Projection to the horizontal equatorial plane. (b) Corresponding stereogram. (After Phillips, 1971.)

practical use this *spherical projection* must be represented in two dimensions. World maps are familiar examples. For present purposes, the most useful way is to project all points on the lower part of the great circle to the horizontal plane by joining them to the zenithal point *P* (Fig. 11.2a), yielding arc *abcd*. Similarly, a structural line through point *O* will intersect the sphere as a point, and this point is projected to the horizontal plane also using *P*; for example, point *b* is the

projection of line *Ob'* (Fig. 11.2a). The resulting representation consists of lines and points plotted inside the horizontal great circle. This limiting circle is termed the *primitive*. This is the method of *stereographic projection*, and the figure drawn on this horizontal diametral plane, together with the cardinal compass directions is a *stereogram* (Fig. 11.2b).

One of the more important properties of the stereographic projection is that a great

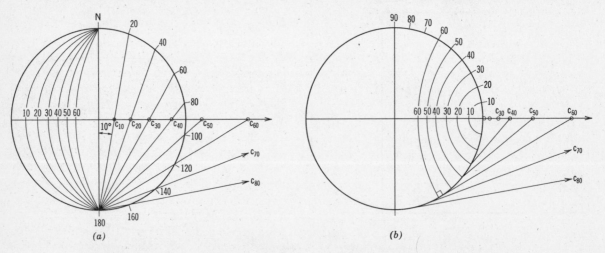

(a)

(b)

FIGURE 11.3 Construction of the stereonet. (a) Centers of the great circle arcs at the intersection of the east-west line and the various chords. (b) Centers of the small circle arcs at the intersection of the north-south line and the tangents to the primitive.

circle on the sphere is also a circle on the stereogram. This permits the representation of any plane to be constructed easily. The geometric centers of the great circle arcs may be found graphically (Fig. 11.3a), or from the relationship:

$$d = r \tan \delta \qquad (11.1)$$

where d is the distance from O to the center, r is the radius of the primitive, and δ is the dip angle. Fig. 11.3a shows a family of meridional great circles representing a series of planes striking due north and inclined to the west at intervals of 10°. Once constructed, a full net of these curves permits the direct plotting of any structural plane.

Planes not passing through the center of the sphere cut the surface as small circles. A second and closely related property of the stereographic projection is that these small circles also plot as circular arcs. These too can be found graphically (Fig. 11.3b), or from

$$d = r/\cos a \qquad (11.2)$$

where, as before, d is the distance from O to the center, r is the radius of the primitive, and a is the angle the small circle makes with a point on the primitive. Thus a family of curves representing a series of such planes can be added to the net (Fig. 11.3b).

The result, in its full form, is the Meridional Stereographic or Wulff Net, or more simply the *stereonet*, in which the two families of curves are drawn every 2° (Fig. 11.4; see p. X-11 for a full-scaled version). The use of this net is a great aid in graphic constructions. Problems are solved by simple manipulation of data which are plotted directly. In short, the net is a portable computer on which many practical problems can be solved quickly, including a number which would be far more taxing by any other manual means. Once the technique is learned, greatest benefit is gained if a net is permanently available. This is easily accomplished if the printed form is mounted on a rigid backing and its surface protected with a clear plastic sheet (see suggested materials, p. viii). In use, data are plotted and problems solved on an overlay sheet of tracing paper. This overlay is affixed to the net by a map pin placed exactly at the center which

allows the sheet to revolve freely. A small piece of clear plastic tape on the back side of the tracing sheet to reinforce the pin hole will prevent tearing or enlarging of the point of rotation.

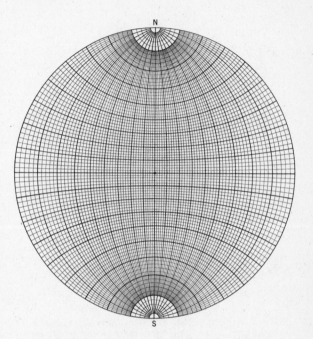

FIGURE 11.4 The Meridional Stereographic or Wulff Net.

TECHNIQUES OF PLOTTING

When using the stereonet, it is important to visualize the net as if you are looking into a hemispheric bowl, and to imagine that the circular arcs are inscribed on its inner surface. Illustrations such as Fig. 11.1b and 11.2a may help in achieving this mental picture. Then the various structural elements to be plotted can be visualized as passing through the center of the sphere and intersecting its surface. *The importance of this visualization can not be overemphasized.* Not only does it make the plotting easier, but it serves as an important check on the proper location, and on the general correctness of the various manipulations. For example, there are four different positions which satisfy the numerical components of dip and strike, but three of these are incorrect. Visualization will quickly show which of these is the correct one. The following examples should be worked through in every detail by the student on his own stereonet. Once the three dimensional picture is clearly and firmly in mind, a variety of short cuts will suggest themselves by which the plotting process can be speeded considerably.

PROBLEM

Given the attitude of a plane (N 30 E, 40 E), plot its great circle representation in stereographic projection.

VISUALIZATION

With the net in front of you (oriented as in Fig. 11.4) hold the flattened left hand, palm upward, over the center of the net with the fingers pointing toward N 30 E, and the plane of the hand inclined 40° to the southeast. The plane of the hand can readily be imagined to extend into the lower hemisphere and intersect its surface (Fig. 11.5a). The trace will cut through the southeast quadrant, and this is where the final plot must also be.

PLOTTING A PLANE (Fig. 11.5)

1. With an overlay sheet in place, make a small mark over the north point of the net and label it *N*.
2. To locate the line of strike, count off 30° clockwise from north, and make a small mark over the primitive at this point.

3. As no great circle on the net passes through this marked point, it is necessary to revolve the overlay until one does. Therefore turn the sheet until the strike mark exactly overlies the north point of the net, that is revolve anticlockwise 30°.
4. To locate the great circle representing a plane dipping 40° east, count off from the primitive on the right side of the net inward along the east-west diameter of the net. Trace in this arc of a great circle.
5. Revolve the overlay back to the original position and check the result by visualization (Fig. 11.5c). Note that it would have been easy to revolve the overlay in the opposite direction, or plot from the left, or both, with erroneous results.

In common with most other projections, the dimensions of the plot are reduced by one. The hemisphere is reduced to a plane, a plane to a line and a line to a point. A further advantage of this particular projection is that a plane can be represented as a point, reducing the dimensions of the plot by one more. For every plane there is a unique line normal to the plane, called the *pole* of the plane. To visualize, hold the hand oriented as before, but with a pencil held between the fingers perpendicular to the plane of the hand. The pencil will pierce the lower hemisphere at a point in the northwest quadrant. This point is everywhere 90°

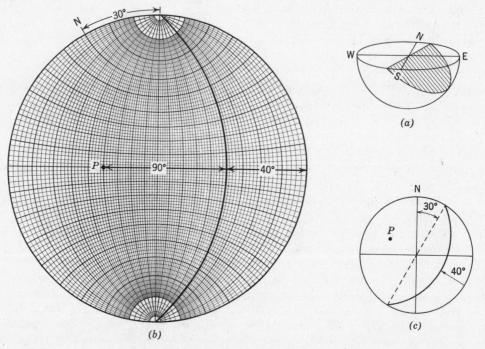

FIGURE 11.5 Stereographic plot of a plane and its pole. (*a*) Perspective view of the inclined plane to be plotted. (*b*) The position of the overlay and net for the actual plot. (*c*) The overlay as it appears after the plot.

from the plane; therefore, from the great circle trace count off 90° from the right to left along the east-west diameter and mark P, the projection of the pole of the plane (Fig. 11.5).

The line which is the pole is projected as a point; this point therefore represents the plane. Any linear structure can be similarly represented by a direct plot, but when a pole is used to represent a plane it is a *reciprocal* plot.

PROBLEM

Given a line (30, S 42 E), plot on the stereonet.

VISUALIZATION

Hold a pencil with the given orientation over the net and visualize its intersection in the southeast quadrant of the hemisphere.

PLOTTING A LINE (Fig. 11.6)

1. With the overlay in place and the south index marked S, locate a point on the primitive representing the trend of the line by counting $42°$ anticlockwise from S.
2. Revolve this trend mark to the south point of the net.
3. Count off 30° from the primitive toward the center along the north-south diameter, and plot the point.

4. Restore the overlay to the starting position and recheck by visualization.

In this particular exercise, the graduations marked by the small circles were used for the first time. However, the trend mark (Step 2 above) could just as easily have been moved to the east point of the net, and the point plotted by counting off along the east-west diameter. In order to assure yourself that this is so, revolve your plotted point to the east-west line and check that the vertical angle measured here is also 30°. Thus in some routines there is a choice of plotting positions. This confuses some beginners, and it is advisable to stick closely with the listed steps until confidence develops. Once the process becomes familiar, however, it will be found that the use of these alternative techniques increases the speed of plotting.

Just as structural lines and planes often occur in combination, so too can they be combined in a single, simple plotting routine.

PROBLEM

Given a plane (N 0, 45 W), and a line in that plane (31, N 36 W), plot both features on the stereonet.

VISUALIZATION

The flattened hand with a pencil held against it in

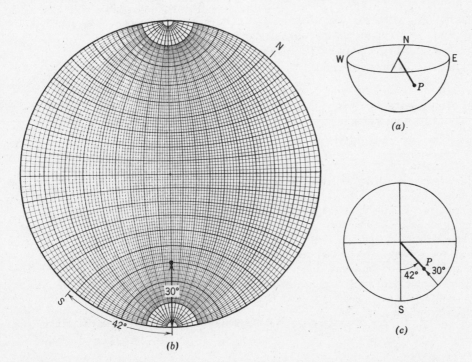

FIGURE 11.6 Stereographic plot of a line. (*a*) Perspective view of the inclined line. (*b*) The position of the overlay and net for the actual plot. (*c*) The overlay as it appears after the plot.

the proper orientation helps one see the three-dimensional aspects of the problem more clearly.

PLOTTING A PLANE CONTAINING A LINE (Fig. 11.7)

1. To plot the plane:
   (a) Mark the direction of strike on the primitive (in this special case, the north mark serves this function).
   (b) Count off 45° from the left on the east-west diameter of the net.
   (c) Trace in the great circle on the overlay (dashed arc on Fig. 11.7).
2. To plot the line:
   (a) With the overlay oriented to north, mark the trend of the line.
   (b) Revolve this mark to the north point of the net. Count off 31° from this point along the north-south diameter.
   (c) Plot the point.
3. Just as the line lies in the plane, so too must the point representing the line lie on the great circle (solid arc of Fig. 11.7). If it does not, then an error has been made, either in plotting or in the original measurement.

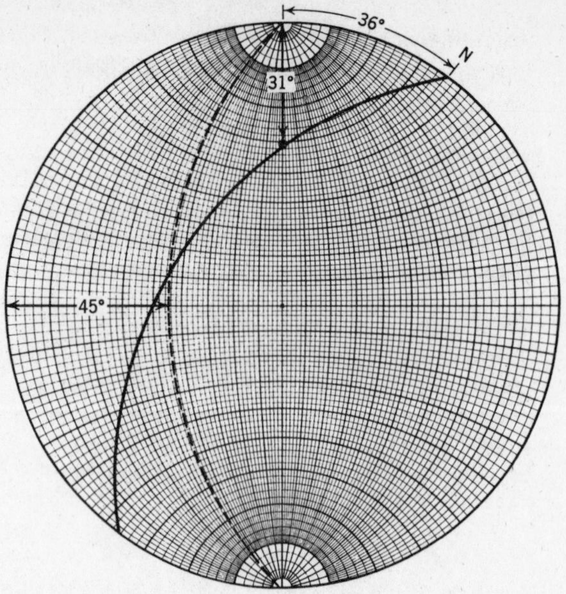

FIGURE 11.7 Stereographic plot of a plane containing a line: the overlay in position for locating the line on the plane.

ATTITUDE PROBLEMS

Problems dealing with angular relationships of planes and lines which were solved by orthographic methods in Chapters 1 and 4 can be solved directly on the stereonet.

PROBLEM

Given an inclined plane (N 50 E, 50 SE), find its apparent dip in the N 80 E direction.

CONSTRUCTION OF APPARENT DIP (Fig. 11.8)

1. To plot the plane:
   (a) Revolve the north mark on the overlay 50° anticlockwise.
   (b) From the east point of the net, count off 50° along the east-west diameter.
   (c) Trace in the great circle.
2. Revolve the overlay back to the starting position, and mark N 80 E on the primitive.
3. Revolve this mark to the east point, and read off the angular position where the great circle crosses the east-west diameter.

ANSWER

The apparent dip in the N 80 E direction is 31°.

From such a diagram, the pitch of a line on an inclined plane can be obtained by determining the angle between the primitive and the point measured *along* the great circle trace. In the example, the pitch of the line is 42° NE (that is, measured from the northeast end of the great circle).

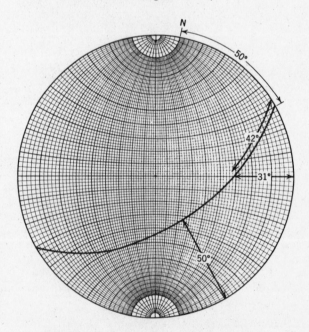

FIGURE 11.8 Apparent dip from true dip and strike: the overlay in position for measuring the apparent dip angle.

PROBLEM

Given two apparent dips, (1) 28, N 56 W, and (2) 22, N 14 E, find the true dip.

CONSTRUCTION OF TRUE DIP (Fig. 11.9)

1. Plot the two apparent dip lines:
   (a) Line 1: revolve the north mark 45° clockwise and count off 30° from north along the north-south diameter.
   (b) Line 2: revolve the north mark 14° anticlockwise and count off 22° from north.
2. Revolve the overlay until the points representing the apparent dips lie on the same great circle. Trace in this arc. The true dip of the plane is read when traced; the strike is easily determined by restoring the overlay to the north position.

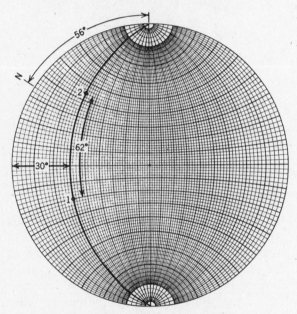

FIGURE 11.9 True dip from two apparent dips: the overlay in position for locating the great circle.

ANSWER

The true attitude is N 56 E, 30 N. The angle between the two lines is the angular distance between the two points (= 62°).

PROBLEM

Given two planes, (1) N 50 E, 60 SE, and (2) N 70 W, 20 S, find the plunge of the line of intersection.

CONSTRUCTION OF THE INTERSECTION OF TWO PLANES (Fig. 11.10a)

1. Plot the two planes:
   (a) Plane 1: revolve the overlay 50° anticlockwise from north and count off 60° from the east point along the east-west diameter. Trace in this great circle.
   (b) Plane 2: revolve the overlay 70° clockwise from north and count off 20° from the west point on the east-west diameter, and complete the great circle.
2. The point of intersection of the two great circles represents the line of intersection of the two planes. To read the plunge angle and bearing, revolve this point until it lies on the north-south diameter of the net.

ANSWER

The plunge of the line of intersection is 20, S 38 W.

Another useful relationship between two intersecting planes is the dihedral angle. This can be easily determined by measuring the angle between the poles of the two planes. Alternatively, by constructing the great circle of which the line of intersection is the pole, the angle between the two planes can be read directly (See Fig. 11.10b). Note that the poles of the planes lie on the great circle perpendicular to the line of intersection.

ROTATIONS

In a number of situations it is necessary to geometrically *rotate* structural elements in space. Every rigid body rotation can be defined by an angle and sense of rotation about a specified axis. The simplest rotation to perform on the stereonet is when the axis R is vertical. Fig. 11.11 illustrates a plane (N 0, 45 E) rotated 45° clockwise about a vertical R to a new orientation (N 45 E, 45 SE). Either the great circle trace or the pole of the plane may be rotated with equivalent results. As is evident from this figure, to find the new position one simply revolves the overlay sheet by the required angle—a familiar manipulation. Yet there is an important difference. Before, the process of turning the overlay about the center of the net was one of convenience in plotting and measuring, but the overlay always carried with it the North mark, so that the original orientations were never really changed. The term *revolve* has been used specifically to describe this maneuver. In contrast, after rotation a plane or line has an entirely new orientation relative to some fixed coordinate direction.

A rotation about a horizontal axis can also be performed readily on the stereonet. First,

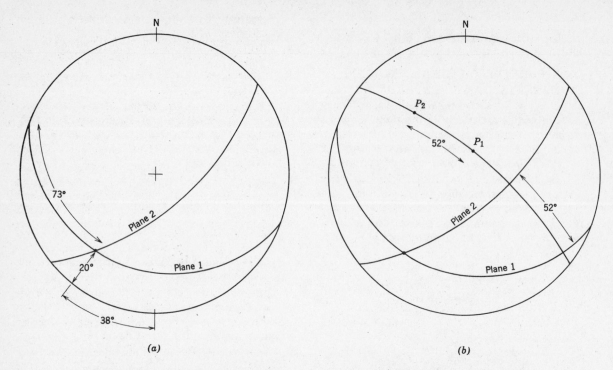

(a)

(b)

FIGURE 11.10 The line of intersection of two planes. (a) The overlap after the plot showing the plunge and pitch of the line. (b) The dihedral angle between the two planes.

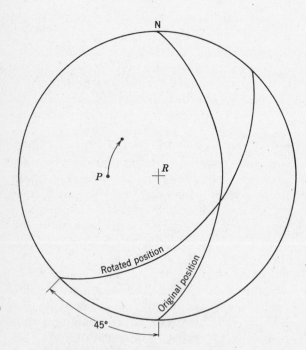

FIGURE 11.11 The rotation of a plane about a vertical axis.

the overlay is revolved so that R coincides with the north-south diameter of the net. In this position, a rotation moves points along the small circle paths. In Fig. 11.12 a plane dipping 60° is rotated anticlockwise as viewed from the south end of R. Although either points or great circles may be rotated, it will be found that working with points is much easier.

It is sometimes necessary to rotate a structural element to horizontal and beyond. Fig. 11.13 illustrates how this is handled. A line (30, N 29 E) is rotated anticlockwise 100°. After just half of this rotation the point lies on the primitive—the line is horizontal. With a further increment of rotation the other end of the line moves into the lower hemisphere at a point diametrically opposite and proceed along the same small circle.

Two methods for rotating about an inclined axis are available. The first depends on previous methods, and consists of rotating R to a horizontal orientation, performing the

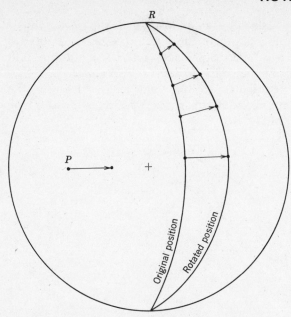

FIGURE 11.12 The rotation of a plane about a horizontal axis.

required rotation, and then returning R to its original position. The second is more direct, though perhaps somewhat more time consuming.

PROBLEM

Rotate the plane (N 83 E, 52 S) 80° clockwise, as viewed looking toward the northeast, about an axis plunging 30° to the N 42 E.

APPROACH

As a pole P rotates about the axis R with constant angle, it will generally describe a small circle on the surface of the sphere. This circle projects as a small circle on the stereonet. While it is useful to draw this circle as an aid to visualization (Fig. 11.14; see Chapter 14 for the method), it is not necessary to do so. A sketch will do. The construction technique consists of rotating the plane containing both the rotational axis and the line in question, rather than rotating the line directly.

CONSTRUCTION (Fig. 11.14; after Turner and Weiss, 1963, p. 69)

1. Plot the rotational axis R and the pole P of the plane to be rotated.
2. Construct the great circle trace representing the plane perpendicular to R.
3. Construct the trace of the plane containing P and R to intersect the plane of step 2 at L. The angle between P and R can be easily read (= 41°).
4. As P rotates about R, so too will the line of intersection L rotate in the plane perpendicular to R. To find the final position of L (= L') count off the required 80° from L going clockwise. In this

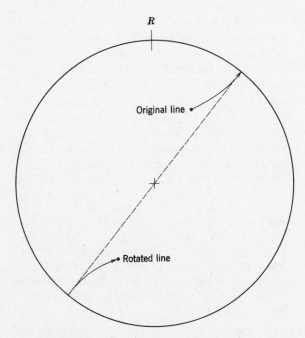

FIGURE 11.13 The rotation of a line to the horizontal and beyond.

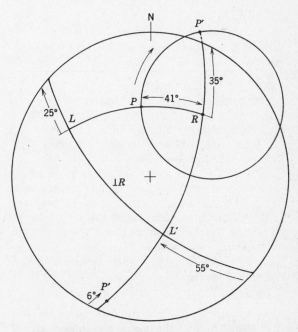

FIGURE 11.14 Rotation of a plane about an inclined axis.

example, the line passes through the primitive so that the 80° is measured in two segments (25° + 55°).

5. Measure 41° from R to P′ in the great circle L′R.

ROTATIONAL PROBLEMS

It is often of interest to determine the orientation of a given feature as it existed before tilting (Fisher, 1938). Simple examples include the restoration for paleogeographic studies of primary sedimentary features such as current lineations (see Potter and Pettijohn, 1963, p. 259), and the pretilt attitude of structures below an angular uncomformity. The most common type of tilting movement occurs during folding, but may also be associated with faulting. With flexural folds, it is a simple matter to unfold the structure and thus restore the beds to a horizontal position. Provided there are no distortions due to strain the various features contained within the folded rocks are thereby also returned to their original positions.

To restore to horizontal the beds of a nonplunging fold, the bedding planes are rotated about an axis parallel to the strike of the beds, which is also parallel to the fold axis, through an angle equal to the dip angle.

PROBLEM

An inclined bed of sandstone (N 20 E, 20 W) contains cross-bedding (N 72 W, 21 S). Determine the original current direction.

CONSTRUCTION (Fig. 11.15)
1. First plot the pole of the cross-beds (=P), and then the sandstone bed as a great circle (=Plane 1).
2. With Plane 1 still in the plotting position, the rotational axis, which is also the line of strike of the sandstone bed, is north-south.
3. To restore the sandstone bed to horizontality Plane 1 rotates 20° to the primitive. At the same time P moves along a small circle in the same direction and by the same amount to P′.
4. From this new pole position P′ the great circle representing the restored cross-bedding can be drawn (=Plane 2). The original current direction is parallel to the dip.

ANSWER

The original attitude of the cross-beds was N 67 E, 30 S, and the associated current moved toward S 24 E. Note that if the orientation of the tilted cross-beds is assumed to reflect the original current direction, an error of 41° is introduced.

The movement leading to the development of a plunging fold can be considered to have two rotational axes: one of them the fold axis, and the other a horizontal axis perpendicular to the fold axis. Reversing the rotation about these two axes unrolls the fold.

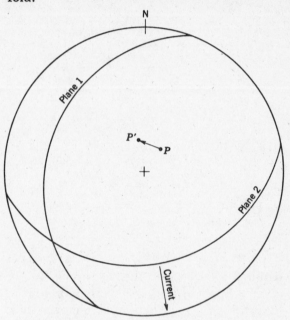

FIGURE 11.15 The two tilt problem.

PROBLEM

Given an anticline plunging 30° due north. Beds on the east limb (N 19 W, 60 E) contain sole markings which trend due east. Determine the original orientation of this sedimentary lineation.

CONSTRUCTION (Fig. 11.16; after Ramsay, 1961)
1. Plot the geometrical elements of the problem: Plane 1 = plane bedding, l = lineation within bedding, and F = fold axis.
2. The rotation of Plane 1 and l about the inclined axis F could be constructed (as in Fig. 11.14), but there is a simpler approach. If the beds are unrolled about the fold axis, the result will be a plane dipping 30° due north. During this rotation, the angle between l and F remains constant. Thus the plane after the first rotation (= Plane 2) and the associated lineation (= l′) can be plotted directly.
3. In rotating Plane 2 about its line of strike to horizontality, the lineation moves along a small circle to the primitive (= l″).

ANSWER

The original trend of the sedimentary lineation was N 65 E. Again, if the correction is ignored, a considerable error results.

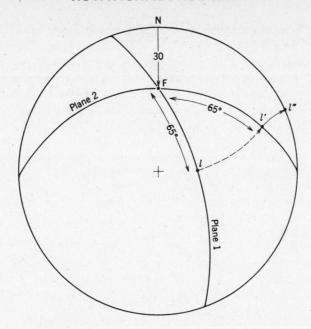

FIGURE 11.16 Unfolding about an in-clined axis.

EXERCISES

1. Construct a stereogram 15 cm in diameter of the 45° great and small circles graphically or with the aid of equations (11.1) and (11.2). Compare your results with the printed Wulff net.

2. Repeat exercises 2 and 3 of Chapter 1. Compare the stereographic and orthographic projection methods for accuracy and speed.

3. Determine the plunge of the line of intersection and the pitch of this line in one of the planes for each of the following pairs of planes:

    a. N 60 W, 46 S; N 15 E, 20 E.   (*Ans.*: Plunge = 17, S 43 E)
    b. N 25 E, 33 W; N 36 W, 70 SW.   (*Ans.*: Pitch in Plane 2 = 35°N)
    c. N 65 W, 50 N; N 25 E, 90 (vertical).

4. A plane contains two linear structures: Line 1 (30, N 40 W) and Line 2 (20, N 10 E). What is the attitude of the plane, and what is the angle between the two lines measured in the plane? (*Ans.*: Plane = N 48 E, 30 NW)

5. The beds below an angular unconformity have an attitude of N 30 W, 40 W. The sequence above the unconformity is tilted (N 20 E, 30 E). What was the attitude of the lower beds before the tilting of the younger beds occurred?

6. An anticlinal fold axis plunges 24, N 40 E. On the east limb, where beds have an attitude of N 5 W, 32 E, the crest line of current ripple marks pitches 70° N in the plane of the bedding. What was the pretilt orientation? Compare the result with that based on the assumption that the tilted lineation adequately represents the original direction. Comment.

7. A fold plunges 50, N 25 E. At a point on the overturned limb, a lineation is found to trend due east, and the strike of the plane containing the lineation is due

north. What was the orientation of the lineation before folding? (This is sufficient information to solve the problem. *Ans.:* N 10 W).

8. Rotate a line (40, N 45 W) 50° anticlockwise (as viewed northward along the rotational axis) about an inclined axis (30, N 20 W). Perform this maneuver in two ways: (1) as a single rotation about the inclined axis, and (2) as a series of steps involving rotation of the axis to horizontal, rotating the line about the axis, and returning the axis to its original orientation.

12

Linear and Planar Structures in Tectonites

GEOMETRIC PROPERTIES OF ROCKS

Homogeneity. If any two identically oriented, equal volume samples taken from a rock mass are identical in every respect, the mass is said to be *homogeneous*. At best some rocks are only quasihomogeneous, that is, the distribution of the various different mineral components is only approximately uniform. Samples from such a mass that are large compared with the grain size will then be statistically indistinguishable, and the mass statistically homogeneous. If a mass is not homogeneous on a given scale, and none are homogeneous on all scales, then it is said to be *hetereogeneous* for that scale.

A mass may also be thought of as being homogeneous with respect to structural properties. A large body of undisturbed sedimentary rocks is said to be structurally homogeneous with respect to bedding, that is, all parts are characterized by a horizontal planar structure. If the same mass is cylindrically folded, it will be homogeneous with respect to fold axes, but hetereogeneous to bedding.

Isotropy. The manner in which the components of a rock body and the boundaries between components are arranged in space is the *fabric* of the body. These components and their boundaries are *elements* of the fabric,

and may be either planar or linear. A rock with randomly oriented fabric elements will have the same physical properties in all directions, and is said to be *isotropic*. Such rocks are rare. Almost all show some degree of preferred orientation, and are therefore *anisotropic*.

TECTONITES

During the deformation of solid rocks, or viscous magma, fabrics develop that are related to patterns of flow, and rocks that show these relationships are called *tectonites*. The most easily studied examples are those in which inequidimensional mineral grains show a preferred form orientation, as in a mica schist. Planar and linear fabric elements may show a variety of planar and linear structures in rocks, either separately or in combination (see Table 12.1).

ANALYSIS OF LINEAR AND PLANAR FABRICS

At the outcrop or in the hand specimen, planar or linear structures are visible as traces on exposure faces. If the structure is simple and well developed, there may be no problem

103

TABLE 12.1   CLASSIFICATION OF LINEAR AND PLANAR FABRICS (after Den Tex, 1954)

1. Planar structures
   a. Planar parallelism of planar fabric elements (Fig. 12.1a)
   b. Planar parallelism of linear fabric elements (Fig. 12.1b)
2. Linear structures
   a. Linear parallelism of linear fabric elements (Fig. 12.2a)
   b. Linear parallelism of planar fabric elements (Fig. 12.2b)
3. Composite structures
   a. Combined structures: two or more planar and/or linear structures in combination.
   b. Complex structures: two or more fabrics marked by either linear or planar fabric elements only.
      (1) Linear + planar fabrics marked by linear elements
      (2) Linear + planar fabrics marked by planar elements

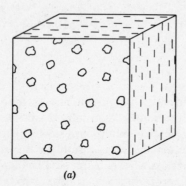
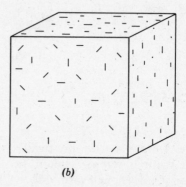

(a)   (b)

FIGURE 12.1   Planar structures: (a) Planar fabric elements, (b) linear fabric elements. (From Oertel, 1962, Bulletin of the Geological Society of America. Used by permission.)

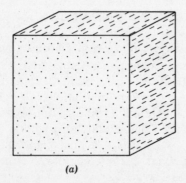
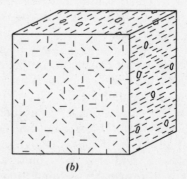

(a)   (b)

FIGURE 12.2   Linear structures.: (a) linear fabric elements,(b planar fabric elements. (From Oertel, 1962, Bulletin of the Geological Society of America. Used by permission.)

in determining its nature and attitude. However, when the traces are faint or several different traces occur on the same face, it may be difficult to tell whether a planar or linear structure is present merely by inspecting several two-dimensional exposure faces. If the traces and the exposure planes on which they occur are measured they can be fitted into a three-dimensional picture on the stereonet.

Planar Structures. If a plane is present, then each trace is an apparent dip, and the method for finding true dip and strike from two apparent dips may be used (Fig. 11.8). However, more than two points are required to demonstrate that the structure, is, in fact, a plane, and the more points that are used the more certain is its existence. With only two points, a plane that satisfies the measurements can be found and this then used to check the result on an exposure face (natural or artificial) perpendicular to the plane.

PROBLEM

The following measurements could have been taken from five different outcrop faces, or from an oriented specimen on which five nonparallel faces had been cut. Determine the attitude of the planar structure.

Exposure plane	Attitude	Pitch of Trace
1	N 40 E, 21 SE	76 S
2	N 75 W, 32 N	20 NW
3	N 80 E, 32 S	76 E
4	N 30 W, 65 W	44 NW
5	N 12 W, 45 W	86 N

CONSTRUCTION (Fig. 12.3)
1. Using the pitch angle, plot the point repre-

senting the measured trace on each exposure face. The great circle of the exposure plane is, of course, used in locating this point, but it need not be added to the diagram.
2. Rotate the points representing all the traces until they fall on the same great circle. This great circle represents the structural plane defined by the measurements.

ANSWER

The plane has an attitude of N 50 W, 55 SW.

Linear Structures. If the structure is linear, traces will be present on all faces except those perpendicular to the direction of the line. As shown in Fig. 12.4, a rod-shaped fabric element cut by any oblique exposure plane P will be exposed as an ellipse. The long axis of this ellipse is an apparent lineation of the true linear structure, that is, the trace marked by the long axis is an orthographic projection of the linear structure onto the exposure face. The true direction of the line lies in the plane N which is normal to the exposure plane and contains the trace. Normal planes can be constructed from measurements on exposure faces; the intersection of any two will ideally fix the attitude of the line. In practice more

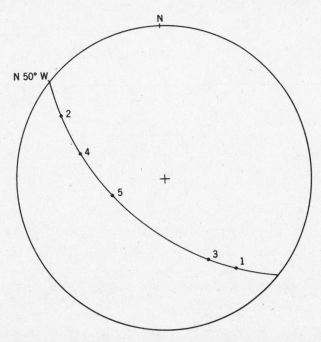

FIGURE 12.3   Attitude of planar structure based on analysis of exposure traces.

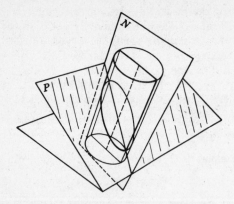

FIGURE 12.4 Relationship between inclined exposure plane *P* and its intersection with a rod-shaped element. Axis of the linear element lies in the plane normal to the exposure plane *N* that contains the trace. (From Lowe, 1946, American Mineralogist. Used by permission.)

points are needed to confirm the nature of the structure and to increase confidence in the result.

PROBLEM

From the following measurements determine the attitude of the linear structure.

Exposure plane	Attitude	Pitch of Trace
1	Horizontal	N 76 W
2	N 12 W, 85 E	45 N
3	N 45 E, 90	44 SW
4	N 30 W, 25 W	48 N
5	N 53 E, 85 SE	34 SW
6	N 46 W, 18 SW	36 NW

CONSTRUCTION (Fig. 12.5*a*; after Lowe, 1946)

1. Plot the points representing the apparent lineation and the pole of each exposure plane.
2. Through each trace point and the pole of the corresponding exposure plane draw great circles.
3. These great circles intersect at point *L* which represents the linear structure.

ANSWER

The attitude of the linear structure is 26, N 76 W.

This method of locating the line of intersection of the *N* planes has an important disadvantage. When the planes intersect at angles of less than about 40° the location of the intersection is influenced by small variations in the measured angles. The variations arise both from measurement errors and natural imperfections in the fabric, and the effect increases as the angle between the

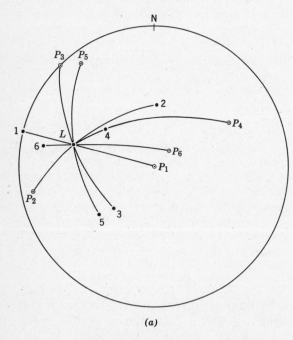

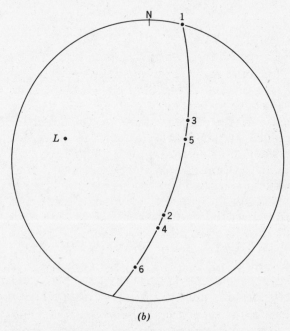

FIGURE 12.5 Attitude of linear structure based on analysis of exposure traces: (*a*) Lowe's method, (*b*) Cruden's method.

planes decreases. In contrast to the idealized example (Fig. 12.5a), this makes it difficult to accurately determine the orientation of the line; in addition concentrations of intersection may appear which are spurious and can not be distinguished from the true lineation. An alternative method of treating the data avoids this problem, and has the additional advantage of being amenable to numerical treatment.

CONSTRUCTION (Fig. 12.5b; after Cruden, 1971)
1. As with Lowe's method (steps 1 and 2), locate the trace points and the poles of the exposure planes.
2. Plot the pole of the great circle defined by each trace-pole pair. These great circles are the same as illustrated in Fig. 12.5a.
3. These new pole points themselves define a great circle which represents the plane perpendicular to the linear structure.

ANSWER
The pole of the best fit great circle found by this method is identical to the intersection L found by Lowe's method.

The techniques for identifying planar and linear structures and determining their attitude may be combined in cases where two or more traces are present on exposure faces.

COMPLEX STRUCTURES

In some rocks, fabric elements of just one shape may be arranged to give more than one structure, even though there is only a single trace on the exposure faces. A simple experiment may help to see how this is possible. A collection of pencils scattered randomly on a table top is analogous to the case of a planar structure marked by linear elements (Class 1b, Table 12.1). Similarly, a parallel alignment of the pencils is analogous to a linear structure marked by linear elements (Class 2a). Now, if these aligned linear elements are slightly dispersed so that all orientations within a small angle of azimuth of, say, 30° are represented, the result is a configuration intermediate between these two end member classes. It possesses both a dominant linear fabric (the still strong alignment) and a subordinate planar fabric (the tendency to

spread in the plane of the table top). A similar pattern involving planar elements can be generated which is intermediate between Classes 1a and 2b.

The analysis of such fabrics depends on the fact that in a given series of random exposures, certain faces will be more favorably oriented for observing the traces of the internal structure than others. Specifically, those exposure planes parallel to a linear structure and those perpendicular to a planar structure will exhibit the best developed traces. Conversely, those planes perpendicular to the linear structure and those parallel to the planar structure will show no traces at all.

PRINCIPAL LINEAR, SUBORDINATE PLANAR STRUCTURE

Such complex structure occurs when, for example, linear elements are statistically arranged with linear parallelism but with a deviation into a plane. The deviation that produces this subordinate structure means that the principal structure cannot be as well developed.

Figure 12.6 shows the stereographic plot of the poles (P) of exposure planes which contained measurable traces (l). Exposure faces oriented so that no trace is visible on them must be examined, but they are not plotted. The measured planes must be well distributed in space for this type of analysis to yield reasonable estimate of the structure.

ANALYSIS
1. The points representing the traces cluster about a center (L) that is the principal linear structure.
2. Note the tendency for the poles of the trace-containing exposure planes to be distributed along a great circle 90° from the linear structure L.
3. The trace points are as well spread out along another great circle (s), which contains six of the 10 l-points, as well as the center L. This is the subordinate planar structure.

Construction of the linear structure using intersecting great circles drawn through the respective poles and traces would have approximated the position L. It would not have detected the spreading that marks the planar structure.

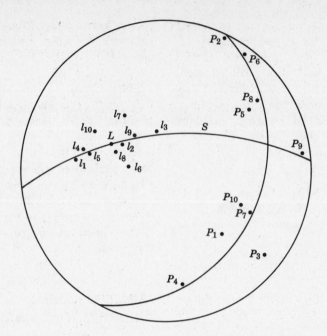

FIGURE 12.6 Principal linear with subordinate planar structure. (From Den Tex, 1954, Journal of the Geological Society of Australia. Used by permission.)

PRINCIPAL PLANAR, SUBORDINATE LINEAR STRUCTURE

A similarly complex structure may result from planar elements the orientation of which marks a planar structure but with deviations that are controlled by a tendency to remain parallel to a line within the plane. Fig. 12.7 shows the plot of such a case.

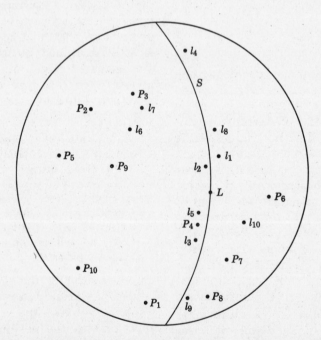

FIGURE 12.7 Principal planar with subordinate linear structure. (From Den Tex, 1954, Journal of the Geological Society of Australia. Used by permission.)

ANALYSIS

1. The majority (7 out of 10) trace points exhibit a tendency to lie close to a great circle (s).
2. There is a similar, though less pronounced, tendency for the poles (P) also to lie along this great circle. Six of 10 poles are less than 45° from s

and are symmetrical with respect to it. The plane s is therefore the principal planar structure.
3. Within s, there is a tendency for the trace points to cluster around a center L. The 4 remaining poles that do not lie close to s are located approximately 90° from L. The point L therefore is a subordinate linear structure lying within the principal plane.

EXERCISES

For the following sets of measurements, analyze the data for the structure involved.

	Exposure	Pitch of Trace 1	Pitch of Trace 2 (when present)
1	N 80 W, 30 N	20 W	
	N 50 E, 80 N	30 W	
	Horizontal	N 46 W	
	N 5 E, 10 E	40 S	
	N 72 E, 20 S	80 W	
2	N 15 W, 70 E	80 N	
	N 52 E, 50 SE	40 NE	
	N 0 , 45 E	74 N	
	N 86 E, 60 S	30 E	
	N 43 E, 50 W	25 NE	
	N 52 E, 35 N	14 NE	
3	N 30 E, 30 W	15 N	85 N
	N 45 W, 20 SW	30 SE	70 SE
	N 20 W, 60 E	52 N	5 S
	N 25 E, 40 E	36 S	28 N
	N 80 W, 70 N	70 E	40 W
	N 50 E, 55 SE	50 SW	15 NE
4	N 70 W, 30 S	28 E	80 W
	N 60 W, 10 S	35 E	90
	N 90 W, 20 N	55 E	73 W
	N 20 W, 40 E	26 N	none
	N 15 W, 45 W	45 S	30 N
	N 55 W, 57 SW	85 S	15 N
	N 50 E, 90	20 SW	60 NE
	N 40 W, 30 NE	60 N	35 N
5	N 50 W, 50 NE	55 NW	
	N 90 W, 30 N	90	
	N 20 W, 50 E	27 N	
	N 10 W, 70 E	40 N	
	N 48 E, 26 NW	46 NE	
	N 20 W, 70 E	68 N	
	N 32 E, 52 W	38 N	
	N 4 E, 70 W	10 N	
	N 85 W, 45 N	90	
	N 85 E, 65 S	none	
	N 20 E, 75 W	60 N	
	N 45 W, 20 NE	46 NW	
	N 90 W, 52 S	none	

13
Structural Analysis

Problems involving the angular relationships of line and planes may also be solved with the methods of descriptive geometry, although the advantages of using the stereographic projection should be obvious. However, if certain problems are to be solved graphically then the use of the stereonet is indispensible. The three-dimensional geometry of a rock mass, especially if complex, is one of these problems. The same basic techniques may also be applied with profit to much simpler situations, and this is a convenient way to introduce the methods.

S-POLE AND BETA DIAGRAMS

In cylindrical folds the hinge zones may be too smooth to allow accurate field measurement, or the folds may be too large or incompletely exposed. If attitudes along the folded surfaces can be measured, the orientation of the fold axis may be determined by a simple plot of the data.

PROBLEM

With the following attitude data, find the fold axis.

1. N 68 E, 30 NW
2. N 60 E, 45 NW
3. N 88 E, 16 N
4. N 35 E, 35 SE
5. N 41 E, 50 SE
6. N 20 E, 20 E

METHODS

There are two different, though equivalent approaches:

1. *Beta diagram*. Plot each measured plane as a great circle. These all intersect at one point, called the β-axis (Fig. 13.1a).
2. *S-pole* (or Pi) *diagram*. Plot the poles of the measured planes. These define a great circle, and the pole of this plane is the β-axis (Fig. 13.1b).

FOLD AXIS AND AXIAL PLANE

The reason for carefully distinguishing between the hinge line and the fold axis may now be appreciated. The β-axis = fold axis in this example characterizes the relationship between any two attitudes, and therefore all attitudes. This axis has no specific location in the fold, only orientation. In cylindrical folds, the hinge lines and the fold axis are parallel, but they refer to quite different aspects of the fold. In simple cylindrical folds, there is a similar relationship between the planar hinge surface of a fold and the axial plane, and there is an interrelationship between both pairs of features, as a simple example will illustrate.

PROBLEM

Given the map of an overturned, plunging anticline, (Fig. 13.2a) we wish to determine the attitude of the fold axis and the axial plane.

CONSTRUCTION (Fig. 13.2b)

1. An *S*-pole diagram of the attitudes around the fold locates the β-axis.
2. With this direction known, a profile may then be constructed to locate the trace of the hinge surface if it has not been found by more direct means. This trace is shown on the map.

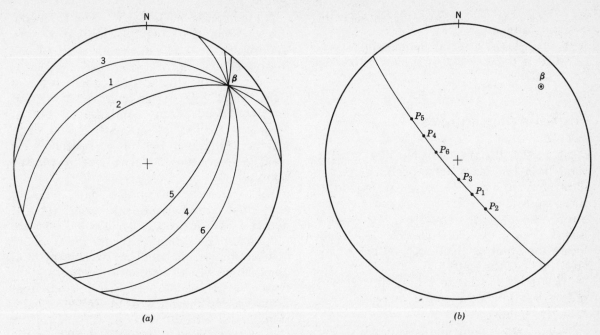

FIGURE 13.1 Stereographic plots of the attitude around a cylindrical fold: (a) beta diagram, (b) S-pole diagram.

3. Add to the stereogram the strike of the planar hinge surface, which is also parallel to the axial plane. As the fold axis is parallel to the axial plane, the axis is, in effect, an apparent dip of that plane. Therefore the great circle through the beta intersection and the strike of the hinge surface gives the dip of the axial plane.

4. For an overturned fold, the orientation of the fold axis may be estimated from the map by inspection (see Turner and Weiss, 1963, p. 166).

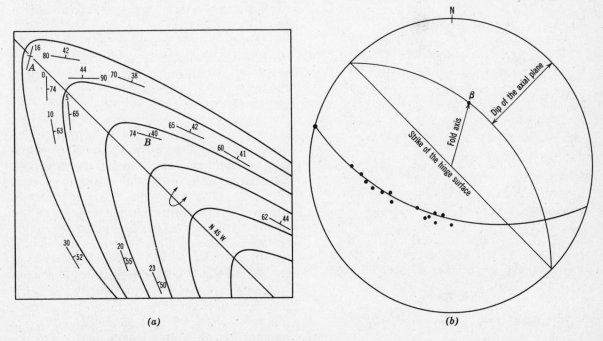

FIGURE 13.2 Attitude of axis and axial plane from map. (a) Map of an overturned plunging fold, (b) Stereogram of axis and axial plane.

a. The axial trend parallels the strike of the vertical attitude (point *A*).

b. The plunge equals the dip of the plane whose strike is perpendicular to the vertical attitude (point *B*).

CONTOURED DIAGRAMS

In practice the stereographic plot of structural lines or planes are never as perfect as illustrated in Fig. 13.1. Irregularities, departures from ideal geometry and measurement errors all contribute to a scatter. If the scatter is small, it is generally possible to visually locate the point or great circle within acceptable limits. If the scatter is greater, it may still be possible to estimate a best fit, but with only a few points the confidence will be low. A larger sample is required.

With a larger number of scattered points, the practical problem of treating and evaluating the data arises. There are several alternatives. Although the theory is not yet fully developed, the statistical evaluation of the orientational data with the aid of a digital computer is certainly destined to be an increasingly important approach (see Watson, 1969; Cruden and Charlesworth, 1972).

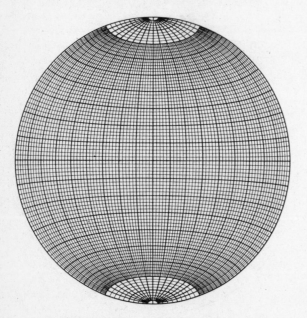

FIGURE 13.3 Schmidt or equal area net.

The computer may also be used to handle the large numbers of data, with the evaluation of printed or plotted output left to the individual (e.g. Spencer and Clabaugh, 1967; Warner, 1969). This latter approach is essentially the older, completely graphical method made efficient by the computer. The most common method of presenting such data, whether processed by the computer or by hand, is to contour the density of the plotted points.

The evaluation of plotted data, whether contoured or not, requires a special type of net. If a series of randomly oriented lines are plotted on the usual Wulff net, the resulting distribution would not be statistically random. There would tend to be a concentration in the center of the net; the random lines would falsely show a weak preferred orientation in the vertical position. The reason for this is that an area (say 10° x 10°) in the center of the net is smaller than the same angular area at the margin. To overcome this, an equal-area or Schmidt net is used (Fig. 13.3). The technique of plotting and manipulating data on this net is identical with that used on the Wulff net. The only practical difference between the two nets is that small circles do not project as circular arcs, and is a problem for certain types of constructions (see Chapter 14).

Once the point diagram is prepared, the densities are counted out. A wide variety of graphical counting methods have been devised (Stauffer, 1966; Denness, 1970, 1972; see also Turner and Weiss, 1963, p. 58f).

Counting out. The method used here is one of the simplest yet devised, and it applies reasonably well to all situations. A special counting net is required which is completely subdivided into small triangles (Fig. 13.4). Six of these triangles form a hexagonal area equal to one percent of the total area of the net. In addition to ease in use, this counting net has the advantage of a fixed relationship between the total number of points and the counted density. Each point is counted three times (except for a small discrepancy caused by the semicircular areas at the ends of the spokes).

FIGURE 13.4 Counting net. (From Kalsbeek, 1963, Neues Jahrbuch für Mineralogie Monatshefte. Used by permission.)

PROCEDURE (Fig. 13.5)

1. Superimpose the point diagram and a second tracing sheet on the counting net. At the center of each hexagon, the total number of points within that hexagon is written (see Fig. 13.5, point A). For

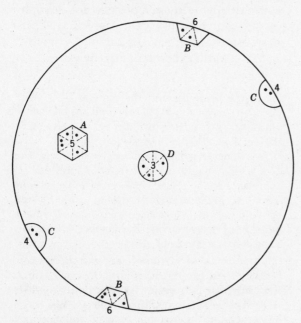

FIGURE 13.5 Counting point densities with the counting net.

the main body of the diagram there will be a number at the center of each overlapping hexagon. For parts of the diagram with no points, the hexagons may be left blank, rather than adding a zero for each.

2. At the periphery of the net, the points in each half hexagon on one side of the net are combined with the complementary half on the opposite side, and this number is written on *both* sides of the net (see point B).

3. Points at the ends of the spokes are counted using the complementary half circles (point C). At the very center the small 1% circle is used (point D).

Contouring. Following the counting out process, the tracing sheet bearing the numerical densities expressed as the number of points per 1% area is removed from the counting net. Contours of equal density are then drawn.

PROCEDURE (Fig. 13.6)

1. To facilitate comparison of diagrams with different numbers of total points, contours are drawn in percentages of the total points per 1% area of the net. Therefore, the number posted during the counting must be converted to percentages. In the special case of exactly 100 points, each number will, of course, also be the required percent figures. If 50 points have been plotted, each point represents 2% of the total, and the posted numbers are doubled, and so forth.

2. Within the main body of the diagram, contours of equal density are drawn as shown at point A (Fig. 13.6*a*). It is usually easiest to locate the area of greatest concentration and work outward.

3. For contour lines that approach the perimeter, the counts along the edge are used. When a contour line intersects the primitive it must reappear exactly 180° opposite (see point B).

4. When a contour line technically should be drawn intersecting the primitive, but it is clear that it loops immediately back again, it is permissible to avoid actual contact (point C).

5. When the preliminary contouring is complete, several modifications may be made in order to improve the diagram (see Fig. 13.6*b*).

a. The maximum found during the counting may not be the true maximum of the diagram. The point of greatest concentration can be found by returning the point diagram to the counting net. Using the central 1% circle, adjust the diagram

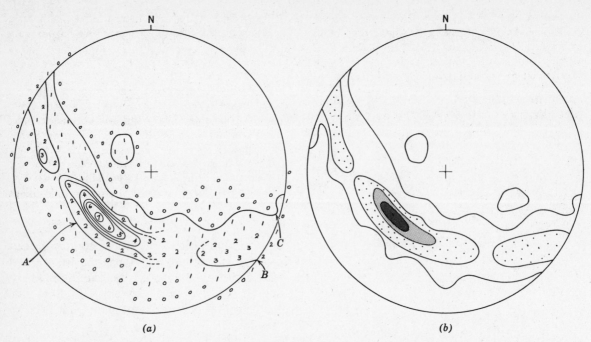

FIGURE 13.6 Contouring. (a) Counted density of 50 points and preliminary contours. (b) Completed contour diagram. Contours 2-4-8-12% per 1% area, maximum 14%.

until the largest number of points lies within this circle.

b. All the contour lines may be unnecessary to show the pattern; for example, if the spacing is very close, and some of the lines may be eliminated. The values of the contours in the final diagram are indicated in the legend in the form 2-4-8-12% per 1% area, maximum 14% (see Fig. 13.6b).

c. The area of maximum concentration is often completely blackened. Although usually unnecessary, patterns may be used for the areas of lesser concentration. Particularly effective are stipple patterns graded so that the areas of greater concentration have a denser appearance. Line patterns detract from the visual effect of the diagram and should be avoided.

INTERPRETATION OF DIAGRAMS

Pattern is the key to interpreting a point diagram and its contoured counterpart. The real equivalents of the perfectly linear and perfectly planar patterns are:

1. the *point maximum*: a symmetrical clustering of points about a single mean orientation.

2. the *girdle*: a grouping of points in a band along a great circle.

For folds, as we have seen, it is possible to choose which of these patterns are to represent the structure (see also Fig. 12.5). There are several compelling reasons for constructing S-pole rather than beta diagrams.

1. In the beta diagram, the number of intersections is equal to $n(n-1)/2$, where n is the number of individual great circle plots. For example, if $n = 25$, the total number of intersections is 300. Such a large number of points is apt to give the impression of a large sample size, and therefore a false sense of confidence in the result. It also involves much more work to produce a beta diagram. For several hundred individual great circle plots, which is not a particularly large sample, the number of intersections becomes impossibly large.

2. As a result of inevitable scatter, spurious concentrations of beta intersections may result. This is especially true in open or tight folds, that is, where angle between the opposing limbs is not large. These spurious intersections will not be randomly distributed about a mean position, and they may exceed in number the significant beta points (Ramsay, 1964).

3. Perhaps the most important advantage is that the S-pole diagram, based on a statistically valid coverage of the structure, gives information concerning the shape of the folded surfaces, the interlimb angle and the attitude of the axial plane.

An instructive approach to understanding S-pole diagrams is to follow the pattern as it

progressively develops during folding. Consider the cylindrical folding of a single layer. Before folding the poles of the horizontal layer would plot as a point maximum at the center of the net (Fig. 13.7a), that is, the poles would define a vertical line. If the diagram were constructed parallel to a vertical

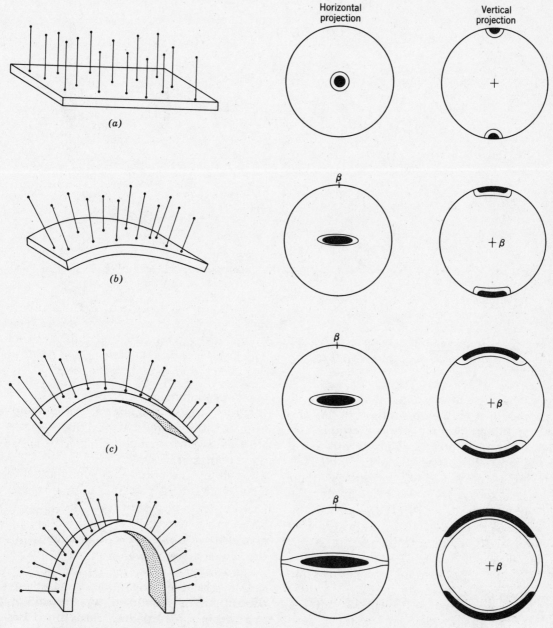

FIGURE 13.7 Development of the S-pole diagram during folding. Note that the same stereographic diagrams would result for both antiforms and synforms: (a) statistically planar horizontal layer, (b) layer bent through 45°, (c) layer bent through 90°, (d) layer bent through 180°.

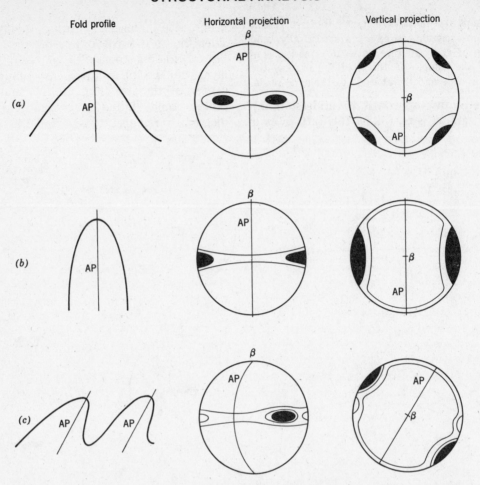

FIGURE 13.8 Patterns of folds: (*a*) symmetrical open fold, (*b*) symmetrical isoclinal fold, (*c*) asymmetrical fold with inclined axial plane.

plane, there would be a point maximum at each end of a diameter of the net. As the layer is folded about a horizontal axis, the originally vertical poles are spread into a fan. In terms of the pattern, whether projected horizontally or vertically, the original point maximum spreads into a *partial* girdle (Fig 13.7*b*). With further folding the girdle continues to spread (Fig. 13.7*c*). Finally, with rotation of the limbs into parallelism, a *full* girdle develops (Fig. 13.7*d*).

If the fold shape is dominated by a semicircular hinge zone (see Fig. 7.3*a*), the density of points within the girdle will be uniform, and the interlimb angle will be the supplement of the angle between the two extreme poles in the girdle. On the other hand, if the fold shape is dominated by planar limbs (see Fig. 7.3), the *S*-pole pattern will consist of a point maximum associated with each limb, and the interlimb angle will be the supplement of the angle between these maxima. Most folds have shapes and patterns between these two extremes.

It will also be noted that symmetrical folds have symmetrical patterns, both in terms of location and concentration of the points (Fig. 13.8*a,b*). Conversely, the patterns of asymmetrical folds are also asymmetrical; for such folds a large number of variations in the patterns are possible. Fig. 13.8*c* illustrates a simple example: the overall shape of the contours are symmetrical, but the point maxima within the girdle have noticably different values; the stronger one marks the dominant limb of the fold.

For purposes of introduction the folds illustrated above are horizontal or upright or both. The axis and axial plane can, of course, have any attitude, and this will be reflected on the diagram. Several plunging and inclined folds are shown in Fig. 13.9.

An additional aspect of contoured diagrams, especially of the point maximum type, is the strength of the pattern, that is, the degree of the preferred orientation. The value of the maximum density of the points is an obvious measure of this strength, but the reliance on a single value, while ignoring the rest of the pattern, is a weakness. A better approach, suggested by Hopwood (1968), takes into account the entire pattern in a simple way. If the contours of equal density are imagined to be lines of equal elevation, then the patterns can be viewed as relief on the inner surface of the hemisphere. This topography can be reassembled into a single conical hill by measuring the area contained within each area bounded by a contour with a planimeter, and constructing a graph of this area against the corresponding elevation to give the slope of the cone. Hopwood found that such a plot closely approximates a straight line. This slope angle can be used to define a coefficient of the degree of preferred orientation; which then facilitates comparisons of different contoured diagrams.

SUPERPOSED FOLDS

The S-pole diagram may also be viewed as a test for the homogeneity of the fold axes in the area being examined. As such, the diagram can be used to decide if, and in what direction a fold profile can be drawn. On the other hand, the pattern may not be interpretable; the scatter may be such that no S-pole girdle is present. Such areas are inhomogeneous with respect to axial directions. This will be the general case in rock masses that have undergone two or more episodes of folding.

The approach in areas of polyphase folding is to seek smaller, homogeneous subdivisions

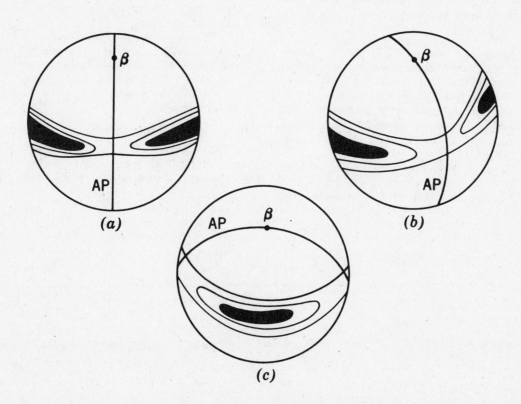

FIGURE 13.9 Folds with different attitudes.

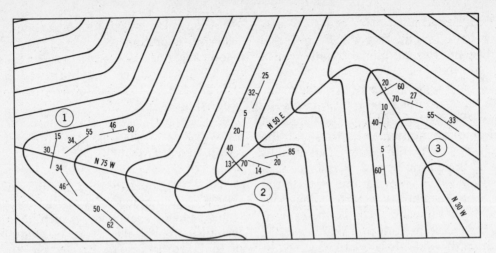

FIGURE 13.10, Idealized map of superposed folds. Subareas 1, 2, and 3 are recognizable by the apparent traces of the hinge surface which are rectilinear.

for which the data does yield interpretable diagrams. A highly artifical example will suggest the approach that is used.

PROBLEM

In an area which has undergone two episodes of folding, determine the geometric relationship between the two sets of folds.

ANALYSIS

1. Subdivide the map area into smaller subareas each of which contains structures that are statistically homogeneous, that is, subareas characterized by cylindrical folds. These subdivisions may be located by trial and error, or by the recognition of rectilinear nature of the apparent traces of the hinge surfaces (Fig. 13.10) or by other structural evidence.

2. Plots of the data from each subarea are then made to determine the orientation of the folds in each homogeneous part of the structure (Fig.

13.11). The changes from one subarea to the next can then be determined by comparing these diagrams.

3. Synoptic diagrams are useful in illustrating these variations, and in obtaining information about the second folds.

　a. Beta intersections of the axial planes from the three subareas define the axis of the second folds (Fig. 13.12a).

　b. The axes of the three subareas lie on a single great circle, which indicates a special type of dispersal of pre-existing fold axes and linear structures during the second deformation (Fig. 13.12b). This pattern of movement is characteristic of similar folding.

In general, results of this type, together with information on the style of folding, permit individual hinge lines to be traced through the superposed folds (Stauffer, 1968).

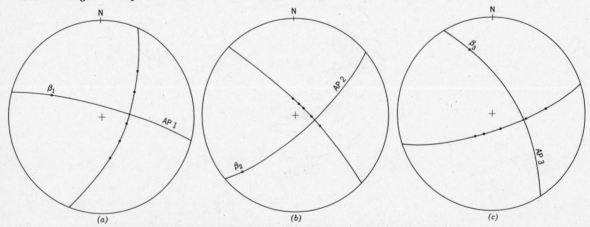

FIGURE 13.11 Stereograms of the data from subareas 1, 2, and 3.

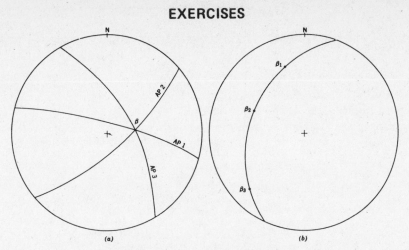

FIGURE 13.12 Synoptic diagrams.

EXERCISES

1. With the attitude data given in Fig. X13.1 construct both a beta diagram and an S-pole diagram. What is the trend and plunge of the fold axes?

2. With Fig. X13.2, determine the orientation of the fold axis by inspection. Confirm with an S-pole diagram.

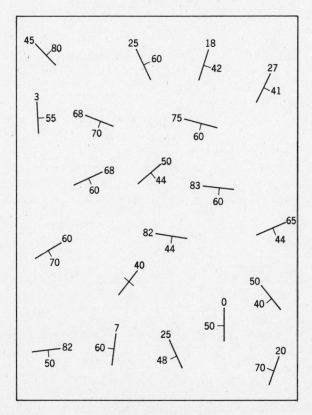

FIGURE X13.1

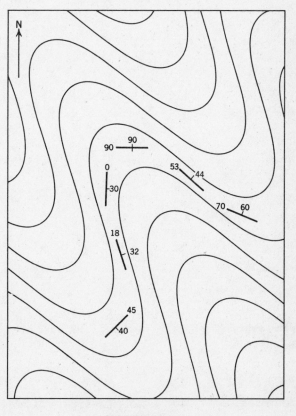

FIGURE X13.2

3. With the data of Fig. X13.3 construct a contoured *S*-pole diagram. Determine the following:
   a. Trend and plunge of the fold axis..
   b. Attitude of the axial plane.
   c. Approximate style of the folds (sketch).
   d. Approximate interlimb angle.
   e. The number of beta intersections if great circles of the attitudes were plotted.

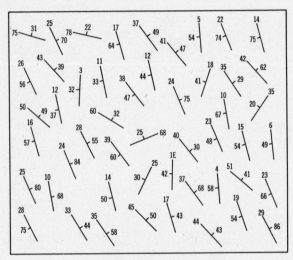

FIGURE X13.3

14

Drill Hole Data

DRILL HOLE DATA

The exploration of the underground by diamond drilling and the recovery of core samples is an important practical technique for the geologist. The information gained from a drilling program depends on the number of holes drilled, the orientation of the holes, and the structures seen in the cores. Because drilling is expensive, it is important to gain the maximum structural data from a minimum number of holes. To plan such a program one must fully understand the geometrical possibilities of one, two and three drill holes.

The problems involved in interpreting drill hole data can be solved in several different ways. Lyons (1964) has given an extended discussion of the mathematical relationships involved, and a series of graphs by which the various equations can be solved. The strictly graphical techniques used here will be introduced with orthographic methods, and then the same problems will be solved much more rapidly on the stereonet.

ONE DRILL HOLE

Ideally, a single recovered drill core could be the equivalent of a measured attitude at a surface outcrop. Although some special devices have been used (Zimmer, 1963), for most drilling it is impossible to keep the core from rotating in the drill pipe during recov-

ery. As a result, only two pieces of information are generally available: (1) the distance along the hole to a recognizable structural plane, and (2) the angle of inclination of the planes in the recovered core with respect to the axis of the drill hole.

The inclination of planes in a core of a vertical hole is the true dip, but because of the rotation of the core during recovery the strike is unknown. We can imagine all the possible original orientations by rotating the core at its original location through 360°. During this rotation the planes in the core will generally describe a right circular cone whose surface widens inward at the dip angle. The intersection of this cone with the earth's surface is a circle (Fig. 14.1a). There are two special cases. If the planes are vertical the cone degenerates to a line, and if they are horizontal it degenerates to a plane. Only in this latter case is the attitude uniquely defined by a single drill hole.

If the single drill hole is inclined, the planes seen in the core do not represent the dip. Again, the true attitude will be tangent to the surface of the cone generated by rotating the core. Depending on the relationship between the drill hole inclination and the angle the planes make in the core, this cone will intersect the earth's surface as an ellipse, parabola, or hyperbola (Fig. 14.1b,c,d). As before, the attitude is completely determined if the planes are perpendicular to the drill core axis.

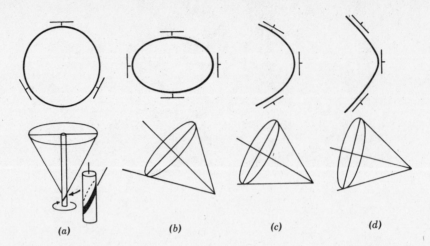

FIGURE 14.1 Cones representing bedding intersected by a drill hole, and the representation of these cones at the earth's surface. (*a*) Vertical drill hole showing the generation of the cone by rotating the core around the axis of the drill hole. The surface trace is a circle. (*b*) An ellipse when the plunge of the hole is greater than the bedding angle. (*c*) A parabola when the plunge and the bedding angle are equal. (*d*) A hyperbola when the plunge of the hole is less than the bedding angle.

From these considerations it will be clear that, in general, the possibilities of strike or dip-strike combinations from a single hole are infinite. More information is needed from additional holes.

TWO DRILL HOLES

A second hole drilled to the same structural plane reduces the possible attitudes to four at most, and in some cases the attitude may be uniquely determined. Several different situations arise depending on the inclination of the holes and on whether a recognizable marker bed is present.

PROBLEM

Two vertical holes are located at the same elevation 350 m apart on an east-west line, and a marker is encountered in the western hole at 100 m and at 250 m in the eastern hole. The bedding makes an angle of 30° with the axis of the core in both holes. What are the possible attitudes?

ORTHOGRAPHIC CONSTRUCTION (Fig. 14.2*a*)

1. Plot the locations of the two drill holes in map view.
2. With the line connecting these two locations as a folding line establish a vertical section and plot on it the drill holes and the depths at which the marker bed would be encountered. The line connecting these two points on the marker defines an apparent dip.
3. At each of the two points on the marker, plot cross sections of the cones formed by rotating the bedding in the core around the core axis. The true dip then appears directly; it could also have been immediately obtained as the complement of the angle the bedding makes with the core axis.
4. The circular intersections of these cones with the earth's surface are then completed in the map view, and the tangents to these circles define the two possible strike directions.

STEREOGRAPHIC CONSTRUCTION (Fig. 14.2*b*)

1. Determine the apparent dip of the marker bed in the vertical section containing the two drill holes, as in Fig. 14.2*a*.
2. Plot this apparent dip on the stereogram as point *A*.
3. Rotate point *A* to each of the two true dip great circles. Each of these then represents a possible true attitude.

ANSWER

The true dip is 60°(=90° -30°). The two strike possibilities are obtained from the tangents in the orthographic construction, or the great circles on

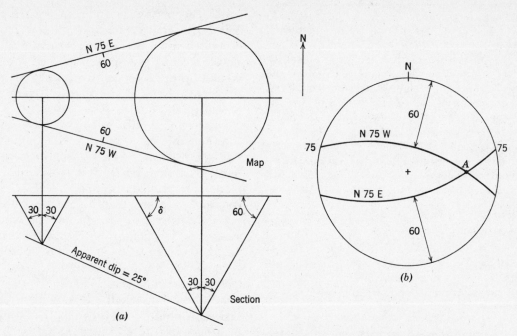

FIGURE 14.2 Construction for two vertical drill holes and a recognizable marker.

the stereogram. Thus the two possible attitudes are N 75 E, 60 S, and N 75 W, 60 N.

PROBLEM

Hole 1 is vertical, and the core-bedding angle is 25°; hole 2 is inclined due west at 45° and the core-bedding angle is 15°. No marker is present. What are the possible attitudes?

CONSTRUCTION (Fig. 14.3)

1. Construct a vertical section showing the inclined drill hole. Without a marker, any vertical hole gives identical information. Therefore a location is arbitrarily chosen so that it intersects the inclined hole at some convenient depth (point D).

2. The intersection of the cone about the vertical

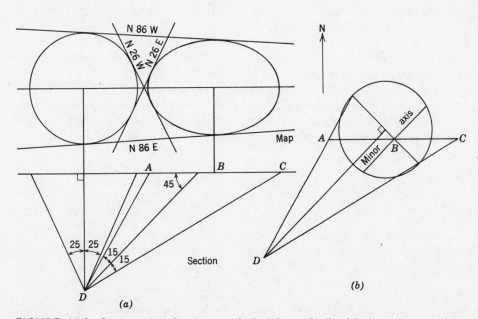

FIGURE 14.3 Construction for one vertical and one inclined hole and no marker bed.

hole with the ground surface is a circle; its location and radius are found directly from the vertical section.

3. The intersection of the inclined cone with the surface is an ellipse. The lengths of the major and minor axes are a prerequisite to constructing this figure. The major axis is *AC*, with center at the midpoint *B* (Fig. 14.3*a*). The length of the minor axis is the "width" of the right circular section passing through *B*. This circle is easily drawn using the cone axis as center (Fig. 14.3*b*). With the length of both axes known the ellipse can be constructed using the method given in Appendix A (p. 187).

ANSWER

The dip angle is obtained directly from the vertical hole; it is 65° (=90°-25°). Four lines can be drawn tangent to both the circle and the ellipse. Each is a possible strike direction, thus four attitudes are possible: N 86 W, 65 S; N 86 E, 65 N; N 26 W, 65 W, N 26 E, 65 W.

This technique of constructing the surface traces of the cones of rotated bedding can be used for all possible drill hole situations. In all cases involving two holes, the tangents common to the two conic sections define the possible attitudes. There are four possibilities; one unique attitude (Fig. 14.4*a*), two (Fig. 14.4*b*), three (Fig. 14.4*c*), and four attitudes (Fig. 14.4*d*). In addition to circles and ellipses, the construction of parabolas and hyperbolas may be required (as in Fig. 14.1), and this presents a problem. The construction of these conic sections may be completely avoided with the use of the stereonet. However, an additional technique is needed.

Instead of constructing the conic section at the ground surface, the intersection of the cone with the lower hemisphere and its representation in stereographic projection are used. The methods of constructing the small and large circle curves on the stereonet is described in an earlier chapter (see Fig. 11.3). These are but examples of one of the fundamental properties of stereographic projection—*any* circle drawn on the sphere will be projected as a circle.

This property will be proved by considering the general case of a small circle (after Phillips, 1963, p.24). A small circle is a section of the surface of a sphere made by a plane not passing through the center of the sphere. It is also the line of intersection of the surface of the sphere and a circular cone with vertex at the center of the sphere. In Fig. 14.5*a*, let line *OV* be the axis of such a cone with vertex angle *LOM*. The intersection of this cone with the surface of the lower hemisphere is a small circle of diameter *LM*, and with point *V* as its center. The projectors of this circle form the cone *LPM*. *LM* is a circular section of this cone oblique to the line *PV*, therefore *LN*, which is perpendicular to *PV*, must be an ellipse. Line *QN* is constructed to be inclined to *PV* at the same angle as LM, and is therefore a conjugate circular section of the cone. *MR* is constructed parallel to the projection plane; then angle *PMR* and *PLM* are equal because they subtend the same arc *PM*. Because *QN* and *LM* are conjugate, angles *PLM* and *PNQ* are equal, and therefore *QN* is parallel to *MR* and the projection plane. Parallel sections of a cone are similar, and therefore the section of the cone *AC* in the projection plane is a circle. Note that since *B*, the projection of the point *V*, is not half way between *A* and *C*, the center of the circle on the sphere is not the center of the circle in projection.

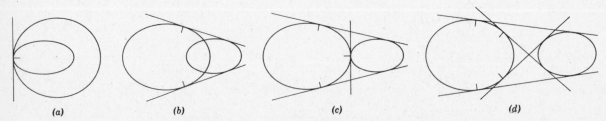

(a)   (b)   (c)   (d)

FIGURE 14.4 All possible two drill hole situations.

It is worth convincing yourself that this actually works in practice, as can be done easily by drawing a circle about the center of the net (see circle A, Fig. 14.5b), and then rotating a number of points on its circumference *and* the center about the north-south axis (circle B, Fig. 14.5b). The center of this new circle is found by bisecting the east-west diameter; it does not coincide with the rotated center. If the circle is rotated further until the original center lies on the primitive (circle C, Fig. 14.5c), it is then in the position of the small circle of the net, and its center can be found by the method of Fig. 11.3b. Any small circle may be constructed directly by first drawing the corresponding cone (section LOM in Fig. 14.5b) and projecting its intersection with the surface of the sphere to the plane of the net to determine the diameter. This method will work even if the small circle lies partially outside the primitive, that is, when the circle extends into the upper hemisphere.

While the graphic method is quite satisfactory for constructing a few small circles, for a larger number it is quicker and easier to calculate r the radius of the small circle, and c the distance from O to its center directly (after Garland, 1971, p. 51):

$$r = \frac{R \sin \Delta}{\sin \phi + \cos \Delta} \tag{14.1}$$

$$c = \frac{R \cos \phi}{\sin \phi + \cos \Delta} \tag{14.2}$$

where R is the radius of the net, ϕ the angle of inclination of the cone axis, and Δ half the vertex angle.

We are now ready to solve all drill hole problems very simply by representing cones on the stereonet as small circles. A simple one drill hole problem will introduce the essence of the method.

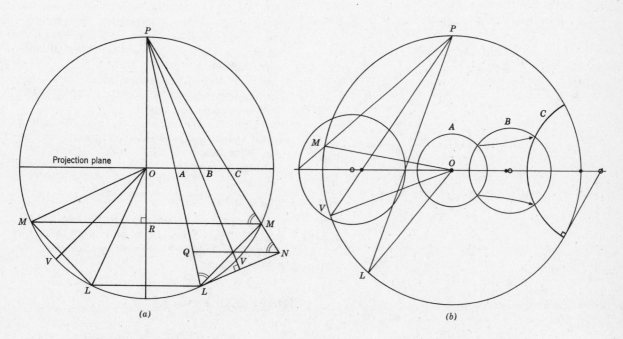

FIGURE 14.5 Small circles on the hemisphere project as small circles. (*a*) The proof and method of construction. (*b*) The rotation of small circle, and the construction of circles which overlap the primitive.

PROBLEM

The attitude of bedding is N 50 E, 50 NW. What will the angle between the bedding and the core axis be for a drill hole inclined 45° due east?

CONSTRUCTION (Fig. 14.6)

1. Plot the pole of bedding P and the drill hole D.
2. Rotate these points to the same great circle and measure the angle between them. It is 36°; the bedding angle is therefore 90° − 36° = 54°.
3. The cone representing the pole of bedding in the core can now be drawn. Count off 36° both ways from D along E-W diameter of the net. Bisect the distance between these points, and with this point as center draw the circle. Note that it passes through point P, as it must.

PROBLEM

Drill hole 1 is inclined 60° due north; it encountered a marker bed at 28.5 m, and the bedding-core angle is 38°. Hole 2, located 100 m due south of No. 1 and at the same elevation, is inclined 50° due west. The marker is found at 52.5 m, and the bedding-core angle is 22°. What is the attitude of the marker bed?

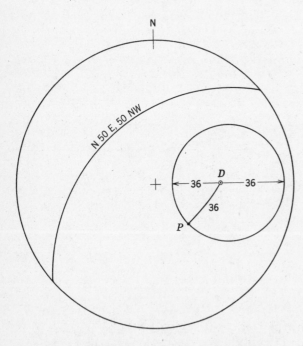

FIGURE 14.6 Prediction of the bedding angle in a drill hole.

STEREOGRAPHIC CONSTRUCTION (Fig. 14.7a)

1. To construct the pole circle representing the bedding in hole 1, plot the drill hole, and count off 52° (=90°-38°) in both directions from this point on the north-south diameter of the net. With the mid point of this diameter as center draw the circle.
2. The small circle associated with hole 2 is constructed in essentially the same manner. In this case, however, it overlaps the primitive. To find the outer end of the diameter of the circle, convert the net to a section, plot the 50° inclination of the cone axis and then count 68° upward. A line through this point and the North point of the net locates the required outer limit of the circle (cf. Fig. 14.5b). The circle is drawn as before.
3. These small circles intersect at two points, each representing a possible pole of the structural plane: A = N 30 E, 30 W and B = N 68 E, 80 S. To choose between these two possibilities, the information supplied by the marker bed must be utilized.

ORTHOGRAPHIC CONSTRUCTION (Fig. 14.7b,c)

1. Plot the locations of the drill holes and their trends in map view (Fig. 14.7b).
2. To determine an apparent dip of the plane the horizontal and vertical locations of the marker bed at two points must be determined. For each hole fold around the surface trace of the inclined hole and plot the plunge angle and the inclined distance to the marker.
3. Locate the surface projection of these intersections with the marker (points X and Y). The depths d_1 and d_2 also appear.
4. Using XY as a folding line, plot the known depths at each end, and thus determine its inclination. With this apparent dip one of the possible attitudes determined on the stereonet can be eliminated.

ANSWER

The apparent dip represented by the line XY is 7, S 27 W. This is compatible only with the plane defined by N 30 E, 30 W, which is therefore the correct attitude.

THREE DRILL HOLES

In any situation three drill holes completely fix the attitude of a structural plane. If a marker is encountered the attitude can be determined by the simple method of the three

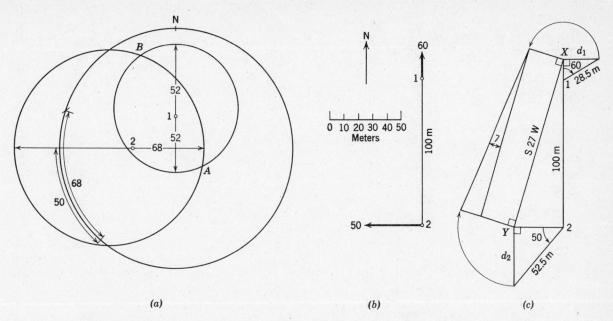

FIGURE 14.7 Construction of attitude from two differently inclined holes and a marker. (*a*) Stereogram giving two possible attitudes. (*b*) Map view of the drill holes. (*c*) Orthographic construction of an apparent dip from data supplied by the marker bed.

point problem. However, its attitude could have been found with only two differently inclined holes. Without marker, the attitude can be determined by finding the unique tangent to three conic sections drawn in map view, or as here, with a direct plot in stereographic projection. A variation of the method of small circles is used to illustrate an even simpler construction which eliminates the need of a compass. Each drill hole is rotated to horizontal and then the overlay is revolved so that the hole is north-south. In this position the traces of the cones may be sketched from the small circles already present on the stereonet.

PROBLEM

Three nonparallel drill holes have the following orientations and core-bedding angles:

1. 60, due north, and 28°
2. 50, N 90 W, and 22°
3. 55, N 45 W, and 25°

Find the attitude of the bedding.

CONSTRUCTION — first pair (Fig. 14.8*a*)

1. Plot the points representing holes 1 and 2.

2. Revolve the net 56° anticlockwise so that these two points lie on the same great circle.

3. Rotate the plane defined be these two points to horizontality; in this position the points (now 1′ and 2′) lie on the primitive. This requires 65° of rotation.

4. Revolve the overlay so that 1′ is on the north pole of the net, and draw the 52° small circle around each pole of the net. Repeat for point 2′ and a 68° small circle. At this point the two small circles may be tangent, or intersect to give two, three or four points (see Fig. 14.9). In this example there are only two intersections A′ and B′.

5. Return the overlay to the position of the great circle of step 2, and by 56° of reverse rotation return the two points of intersection A and B to their true orientation. This gives the poles of two possible attitudes: A = N 30 E, 30 W, and B = N 68 E, 80 S (compare with Fig. 14.7*a*).

The procedure for solving the second pair of drill holes is identical. However, it should be apparent that the complete small circles are not needed to locate the points of intersection, and thus the actual construction may be shortened.

CONSTRUCTION — second pair (Fig. 14.8*b*)

1. Plot the holes 2 and 3.

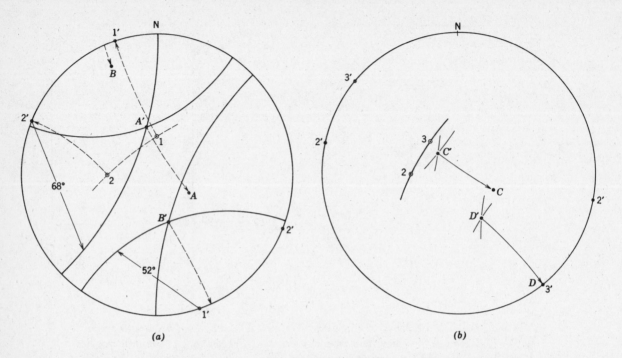

FIGURE 14.8 Construction of attitude from three drill holes. (a) Stereogram of holes 1 and 2. (b) Stereogram of holes 2 and 3, using short cuts.

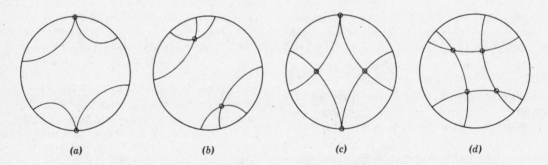

FIGURE 14.9 The four possible two drill hole situations on stereograms. (Compare corresponding parts of Fig. 14.4.)

2. Find the plane containing both.
3. Rotate this plane to horizontality, giving points 2′ and 3′.
4. Draw arcs of the small circle 68° and 65° respectively about these two points on the primitive, giving intersections C′ and D′.
5. Return these intersections to their true orientation by reverse rotation, giving C and D, the two

possible attitudes based on this pair of holes.

ANSWER

Comparing attitudes A and B, with C and D indicates that the true attitude of the bedding planes intercepted by the drill holes is N 30 E, 30 W.

EXERCISES

1. Vertical drill hole No. 1 encountered a recognizable horizon at a depth of 65 m.
   Vertical drill hole No. 2 located 120 m to the N 30 W, encountered the same

marker at a depth of 33 m. The angle between the drill hole axis and the bedding was 40°. What is the dip, and what are the strike possibilities?

2. In a vertical drill hole the angle bedding makes with the core axis is 20°. In a second hole (50, N 45 E), the angle is 15°. What are the attitude possibilities?

3. A vertical hole encountered a marker at 14.8 m, and the core-bedding angle is 60°. A second hole (30, N 20 E) is located 60 m due east. The marker was found at 33.6 m along the drill hole and it made an angle of 45° with the core axis. What is the attitude?

4. Three drill holes intersect rocks with a prominant planar structure. From the following information, what is the attitude of this structure?

Hole	Attitude	Core-bedding angle
1	82, S 37 W	17
2	61, S 21 E	34
3	50, S 7 E	30

5. With a recognizable marker, why do two differently inclined holes give more information than two parallel holes?

15
Faults

DEFINITIONS

Fault

A fracture surface or zone along which appreciable displacement has taken place.

Separation

The perpendicular distance between the two traces of a displaced marker plane measured in the plane of the fault. Components of separation may be measured in a number of specified directions. Two are in particularly common usage: strike separation (measured parallel to the strike of the fault), and dip separation (measured parallel to the dip of the fault); see Fig. 15.1a. Another useful component is stratigraphic separation, which is measured perpendicular to the displaced planes, not in the plane of the fault.

Slip

The relative movement on the fault, measured from one block to the other, as the displacement of formerly adjacent points (Fig. 15.1b).

Foot wall

The lower side of a horizontal or inclined rock body or fault.

Hanging wall

The upper side of a horizontal or inclined rock body or fault.

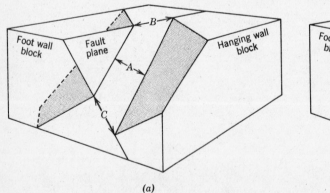
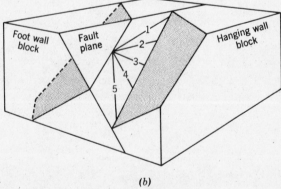

(a)   (b)

FIGURE 15.1 Fault displacing a marker plane. (a) Separation A, strike separation B, and dip separation C. (b) Different orientations of slip: 1 = reverse left-slip, 2 = left-slip, 3 = normal left-slip, 4 = normal-slip, 5 = normal right-slip. (After Hill, 1963.)

130

FAULT CLASSIFICATION

There are several geometrical approaches to the classification of faults. For example, there are descriptive schemes based on the relationship of the fault to other structures (longitudinal, transverse, or bedding plane faults), and on the patterns of groups of faults (radial, parallel, or en échelon faults).

The most important aspect of fault geometry is the relative displacement along it. Slip is the measure of this displacement, but unfortunately it is only occasionally ascertained. Separation, or a component of it, is more commonly measureable, and in fact, such separation is usually the evidence for the existence of the fault. However, a clear distinction must be maintained between slip and separation, because an observed separation may result from many possible orientations of slip (Fig. 15.1b). In order to emphasize the importance of this distinction, two parallel classifications of fault displacements have evolved; one based on slip and the other on separation (Hill, 1963). However the separation-based scheme is not really a classification of displacement at all. This will be clear when it is realized that for a given slip, the amount and sense of separation depends on the attitude of the disrupted marker plane, and two differently oriented planes may show opposed senses of apparent displacement (Fig. 15.2). Thus any classification based on separation is misleading and should be dropped (see Gill, 1971). If the separation on a fault is to be described, it should be spelled out in terms of the attitude of both the attitude of the disrupted plane and the direction in which the separation is measured. The sense of the separation can be described by the use of several commonly used terms (see Table 15.1).

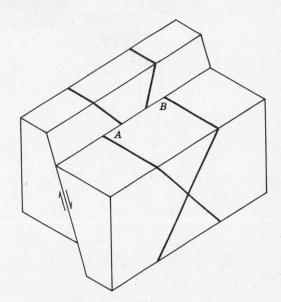

FIGURE 15.2 Fault with opposed senses of strike separation.

The down-structure method of viewing a map allows one to see directly the geometric relationships between slip and separation. In Fig. 15.3, three faults displace inclined strata. Adopting a down-dip view of the beds (not the faults) reveals the thickness of the beds perpendicular to the line of sight. The separation of the beds in this same direction can also be seen; this is the stratigraphic separation. Fault I has a stratigraphic separation

TABLE 15.1   Terms Describing the Sense of Separation

Strike Separation	
Left-separation	Standing on the trace of a displaced marker on one block, the trace of the same marker is found to the left (the displaced plane in Fig. 15.1 shows left-separation).
Right-separation	The trace of the marker is found to the right across the fault.
Dip Separation	
Normal-separation	Viewing the fault in a vertical section, the trace of the marker on the hanging wall block is found below the trace of the same marker on the foot wall block (the displaced plane in Fig. 15.1 shows normal-separation).
Reverse-separation	The trace of the marker in the hanging wall block is found above the trace in the foot wall block.

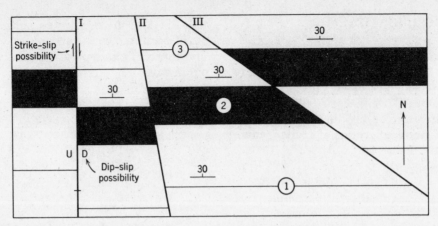

FIGURE 15.3 Diagramatic map illustrating the down-structure method as applied to faults. (After Mackin, 1950.)

equal to the thickness of Bed 2. As an aid to this visualization, it is useful, especially for beginners, to hold the flattened hands, fingers in the dip direction, to represent the two sides of the fault. Moving the hands then reproduces the various slip possibilities. For example, it is easy to imagine a slip which would account directly for the stratigraphic separation—the west block relatively upward and northward at an angle of 60°. It can then be seen that many other slips are also possible. The west block could have moved upward by slip parallel to the dip of the fault, or it could have have moved northward parallel to the strike, or any combinations of these. For Faults II and III the stratigraphic separation is also readily seen regardless of the angle between the strike of the fault and the strike of the beds. Further, the down-dip view reveals the nature of the separation regardless of the dip of the fault.

Clearly *slip* fundamentally describes the displacement on the fault, and a classification based on it is the only meaningful way of categorizing this information. For completeness, it is also necessary to distinguish between *translational* faults, where the amounts and orientation of the slip is everywhere the same, and *rotational* faults, were the slip changes from place to place (Fig. 15.4). Probably every translation fault also has a rotational component, especially near its ends. Table 15.2 gives the full slip-based classification of faults.

The attitude of the fault plane is the second important factor in describing the relative displacement of the two fault-bounded blocks. For example, a reverse-slip fault with a dip of 20° is considerably different from a reverse-slip fault with a dip of 70° It is therefore necessary to introduce dip into the classification of fault displacements. In fact, it may be observed that dip already has entered the slip-based scheme: normal and reverse faults must be inclined.

Rickard (1972) has suggested a useful way of combining the dip angle with the pitch of

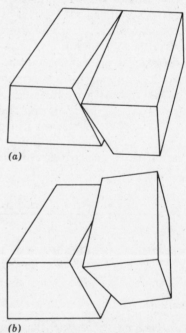

FIGURE 15.4 Rotational faults: (*a*) hinge fault, (*b*) pivotal fault. (From Donath, 1962, Bulletin of the Geological Society of America. Used by permission.)

TABLE 15.2 Classification of Faults Based on Slip (after Hill, 1963; Gill, 1971)

Translational Movement

Dip-slip	Normal-slip fault (hanging wall block down).
	Reverse-slip fault (hanging wall block up).
	For vertical faults, specify the movement of one block relative to the other, for example vertical dip-slip fault, east side down.
Strike-slip	Right-slip fault (opposite block to the right).
	Left-slip fault (opposite block to the left).
	For horizontal faults, describe the direction of movement of the hanging wall block, for example, horizontal north-east slip fault.
Oblique-slip	Dip- and strike-slip terms in combination (Fig. 15.1b), for example, normal right-slip fault, or, for a vertical fault, vertical oblique-slip fault, north side down and westward.

Rotational Movement

Plane fault*	Clockwise-rotational fault (opposite block clockwise).
	Anticlockwise-rotational fault (opposite block anticlockwise).

*Slip on curved fault surface may also occur; see discussion in text.

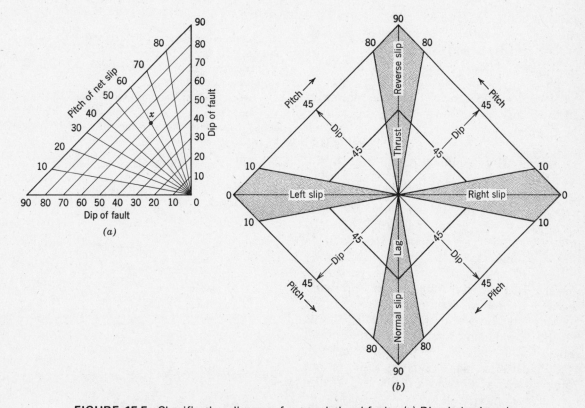

FIGURE 15.5 Classification diagrams for translational faults. (*a*) Dip-pitch triangular grid. (*b*) Fault categories on the dip-pitch triangles. The special cases of dip slip and strike slip faults are shown shaded; the oblique slip catagories are blank. (After Rickard, 1972.)

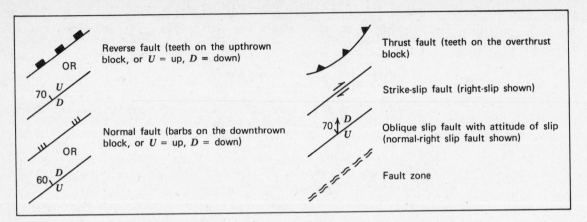

FIGURE 15.6 Suggested fault map symbols. (In part after Hill, 1960.)

the net slip on a triangular diagram, which leads to a graphical classification scheme. Each possible dip-pitch pair can be assigned an index symbol; for example, the symbol for a fault dipping 60° on which the line of the net slip pitches 60° would be $D_{60}R_{60}$ (R, for rake, is used to avoid confusion with plunge). This is then represented by a point on the triangular grid (see point x Fig. 15.5a, cf. Fig. 7.9). Four triangles are necessary to represent normal, reverse, and right and left slips (Fig. 15.5b). In this way all possible translational faults can be plotted. Further, the main categories of faults can be added to the full diagram as an aid to classification. Rickard also suggests that the special cases of dip and strike slip be restricted to pitches of 80-90° and 0-10°, respectively. By including the dip angle in the classification, several additional catagories are needed. A fault dipping less than 45° may be called a *thrust* if the slip is reverse, and *lag* is suggested if slip is normal. Although not precisely defined, an *overthrust* is a low angle thrust. Dennis (1967, p. 157) advises against both thrust and overthrust on the grounds of unwarranted genetic implications. However, there is a need to make this distinction in attitude, and it seems unlikely that the terms will be abandoned. With care geometric and genetic connotations can be separated. Fig. 15.6 illustrates many symbols for several combinations of slip and dip which cover the most common types of faults.

TRANSLATIONAL FAULTS

The determination of the amount and orientation of slip requires the recognition of two originally adjacent points on the fault surface. Strictly, geologic examples of such points are nonexistent, and therefore, other features from which points may be derived must be found. In practice, lines can be recognized in several situations. These lines pierce the fault plane to give the required points (Fig. 15.7). Such lines may be represented by physical structures:

1. Intersecting planes (for example, the intersection of two dikes).
2. Trace of one plane on another (beds truncated against an unconformity).
3. Linear geologic bodies (shoestring sands, linear ore veins, etc.).
4. Stratigraphic lines (pinch-out lines, ancient shore lines, etc.)

or may be constructed from field data:

1. Isopachous or isochore lines.
2. Lithofacies lines.
3. Hinge lines of folds.

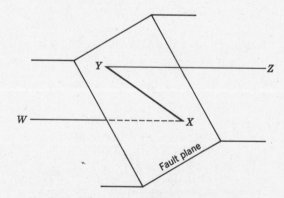

FIGURE 15.7 Piercing points: X = piercing point of line WX; Y = piercing point of line YZ; XY = slip. (After Crowell, 1959.)

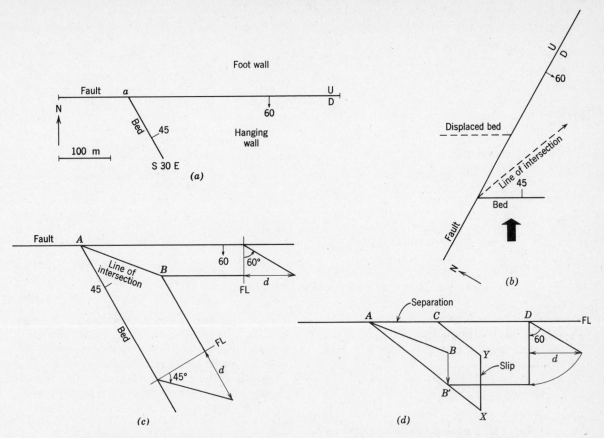

FIGURE 15.8 Prediction of the location of a displaced plane across a fault. (a) Map with the elements of the problem. (b) Down-dip view of displaced plane. (c) Orthographic construction of line of intersection. (d) Determination of strike separation.

PROBLEM

A normal-slip fault displaces a marker bed which is exposed only in the hanging wall block (Fig. 15.8a). If the net slip is 100 m, find the location of the bed on the opposite side of the fault.

APPROACH

Before starting any construction, it is advisable to thoroughly visualize the elements of the problem. This will save time, and help avoid errors. Turn the map so that a down-dip view of the bed is obtained (Fig. 15.8b is in this position). Look down the dip, and at the same time hold the hands together to represent the dipping bed. Now move the left hand upward in the direction of the known slip. It will then be clear that the displaced bed is to be found to the east of its position in the hanging wall block, that is, it must show right-separation. With the map in this same position, the line of intersection between the fault and the bed should also be visualized. In order to determine the exact location of the displaced bed, a view of the fault plane is required.

CONSTRUCTION by orthographic methods

1. To find the line of intersection (Fig. 15.8c):
    a. Establish folding lines perpendicular to both the fault trace and the trace of the bed, and draw structure contours at depth d (this construction is identical to that used in Chapter 4).
    b. The intersection of these contours fixes a second point on the line of intersection: therefore AB is the projection of this line in map view.
2. To draw a view of the fault (Fig. 15.8d):
    a. Point B lies on the d structure contour drawn on the fault plane. Using the fault trace as a FL, and point D as center, an arc locates the slope distance between the fault trace and the d contour.
    b. At the same time, point B moves to B', and AB' is then the line of intersection on the fault plane.
3. To find the location of the displaced bed:
    a. Arbitrarily locate point X on AB' or its extension. This point must have been originally adjacent to some point Y, which is found by

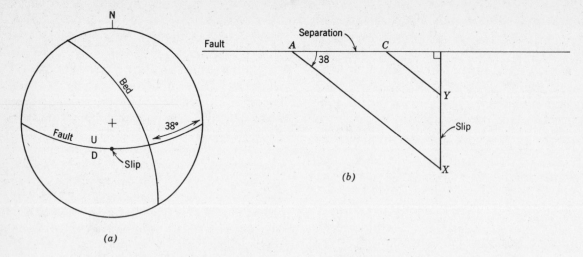

FIGURE 15.9 Location of displaced plane. (a) Stereoplot of the problem. (b) Completion of problem by orthographic construction (cf. Fig. 15.8d).

measuring 100 m in the slip direction parallel to the dip direction of the fault. (Remember that the fault was originally specified as being normal-slip.)

b. Draw YC parallel to AB′. Distance AC is the required separation.

ANSWER

The measured separation is 125 m. With this, the location of the bed on the foot wall side can be found (see Fig. 15.9b).

In normal practice, the steps illustrated in Fig. 15.8c,d would be superimposed on the map (Fig. 15.8a). Even so, it is still a time consuming construction. Use of the stereonet by-passes all but the final steps.

CONSTRUCTION with the aid of the stereonet

1. Plot both the fault and bed on the stereonet, and determine the pitch of the trace of the bed in the plane of the fault (= 38 E in Fig. 15.9a).

2. With this, plot the line of intersection directly on the map. The slip is measured and the separation found exactly as in step 3 above (Fig. 15.9b; cf. Fig. 15.8d).

PROBLEM

Given the east-west fault, dipping 55 S and displacing a bed and a vein shown in Fig. 15.10a, determine the amount and orientation of the slip, and classify the fault accordingly.

APPROACH

Visualize. Turn the map so that a down-dip view of the vein is obtained and with the aid of the

hands, note that both normal-slip and left-slip satisfies the stratigraphic separation of the vein. Repeat for the marker bed, with similar results. The opposed senses of strike-separation rules out the possibility of a significant strike-slip displacement. A strong component of dip-slip is indicated. Again, to calculate the slip a constructed view of the fault plane is required.

CONSTRUCTION

1. A stereogram of the various elements yields the pitch of the bed (30 E) and the vein (50 W) in the plane of the fault (Fig. 15.10b).

2. Using the pitch angles, the traces of the displaced planes on the fault can be drawn, giving the piercing point from the foot wall side X, and from the hanging wall side Y (Fig. 15.10c).

3. The line XY is the slip. The pitch of the slip line is measured from the trace of the fault.

ANSWER

The net slip is 25 m, and the pitch of the slip line on the fault is 82 E. Plotting the slip line on the stereogram allows the full orientation of the slip line to be read: 54, S 13 E. With this, the full attitude symbol can be added to the map. It is a normal-slip fault, with a very small left-slip component.

FAULT DRAG FOLDS

Faults and folds are often found together. There are two distinctly different relationships. First, the association may be only coincidental; folds existed in the rock mass before faulting. Certain geometrical relationships between the preexisting folds and the faults may be detectable, but only to the extent that the folds contribute to the overall mechanical properties of the faulted mass. Second, there may be a genetic relationship between folding and faulting. Two cases are possible:

1. The fault may result from the folding, being a direct continuation of the deformation. A common example is a thrust along the lower limb of an overturned fold. The décollement (Chapter 6) also falls into this class of fold-related faults.

2. The second case, and the one of concern here, are folds that result directly from faulting. During slippage of one block past the other, frictional drag on the fault plane may produce certain effects in the blocks themselves. One of these is the dragging of preexisting layers into folds. Such folds are termed *fault drag folds*. Where present, they represent displacement along the fault zone in addition to the slip. The total displacement, measured outside the zone of disturbance associated with the fault is termed *shift*.

The presence of fault drag folds gives slightly more information concerning the fault movement than in cases where they are absent; the sense of fault separation can be determined from observations made on only one side of the fault. However, great caution is necessary. The presence of fault drag fold tends to encourage an impulse to read into the pattern something more than the limited measurement of separation, that is, to use the curvature of the folds as evidence of the direction of slip. For example, the map pattern of Fig. 15.11a has been mistakenly used as an indication of strike-slip displacement. While this is certainly one of the possibilities, there is no more evidence of such slip than in the map pattern of Fig. 15.11b, where the drag folds are absent.

Two considerations will make this limitation clear. First, a vertical section drawn perpendicular to the fault trace will always show these same folds (see section AA' of Fig. 15.11a). In fact, every section that intersects the fault, and the beds will reveal

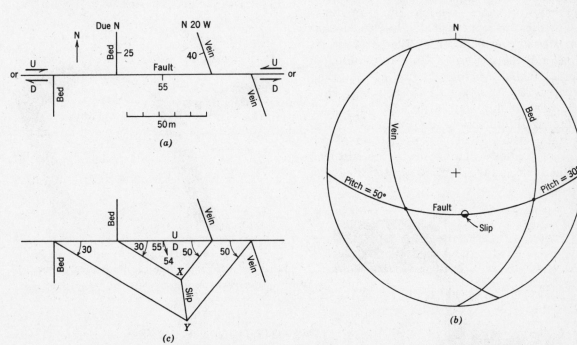

FIGURE 15.10 Slip from two displaced planes. (*a*) Map showing geometry of displaced planes, together with the possible senses of slip obtained from visualization. (*b*) Stereogram showing pitch of the bed and the vein on the fault plane. The slip line is shown by the small split circle with the blackened half indicating the down side. (*c*) Orthographic construction of the slip line.

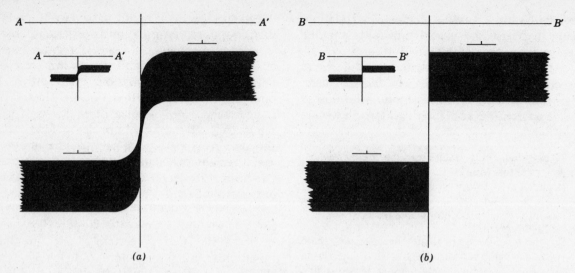

(a)   (b)

FIGURE 15.11 Examples of fault separation: (a) with fault drag folds, (b) without faults drag folds.

curvature. Clearly, the folds of each of these diversly oriented sections can not be taken as evidence of slip parallel to the plane of the section. Thus only separation can be determined—strike-separation in map view and dip-separation in vertical section. The geometry is essentially the same as if the drag folds were not present (see section BB', Fig. 15.11b).

Second, the process of fault drag folding is closely related to distributed faulting (Fig. 15.12). This type of slicing is suggestive of similar folding, and the same limitations on deducing the direction of flow holds. It will then be evident that the orientation of the fault drag folds is determined by the attitudes of the fault and the displaced layers; the hinge

lines are parallel to the intersection of these two planes. A down-dip of the folded beds should amply confirm this (see also Fig. 15.13).

FIGURE 15.13 The axis of fault drag folds is parallel to the line of intersection of the fault plane and the plane of the bedding.

ROTATIONAL FAULTS

There are several different situations in which relative rotation of two fault-bounded blocks may be accommodated. The fault may be cylindrical (Fig. 15.14); in terms of motion this can be more simply considered a case of

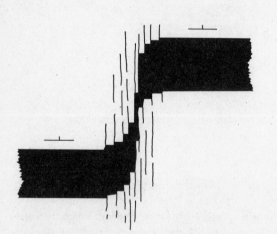

FIGURE 15.12 A distributive fault.

translation on a curved surface. Gill (1971) notes that the rotational axis may be inclined to the fault, for example, where displacement occurs on a conical surface, but this is of uncertain importance. We will confine our attention here to problems involving rotation about an axis perpendicular to the fault plane.

FIGURE 15.14 Cylindrical fault with axis of rotation parallel to the fault surface.

PROBLEM

Determine the angle and sense of rotation of the fault shown in Fig. 15.15a.

METHOD

The traces of the displaced plane from the foot and hanging wall sides can be plotted from pitch angles determined on the stereonet, or more simply, the difference in pitch angles may be read directly.

ANSWER

It is a clockwise-rotational fault, and the rotational angle is 30°.

If two piercing points can be located (X and Y), the center of rotation can also be found — it lies at the apex of the isosceles triangle whose base is the straight line XY and opposite angle is the rotational angle (see Fig. 15.15b). If rotation and translation are combined, this construction will always result in the location of a center which would account for the motion by rotation alone. Thus if a rotational component is present, and only two piercing points are known, the slip is indeterminate. Despite this limitation, the angle of rotation and two originally adjacent points still allows the position and location of all other structures to be predicted from one block to the other. If sets of points can be found at two different locations, and the angle of rotation is constant, the rotational centers will coincide in the case of rotation only, but will differ if a component of translation is involved. It is possible to separate the two components by a trial and error process.

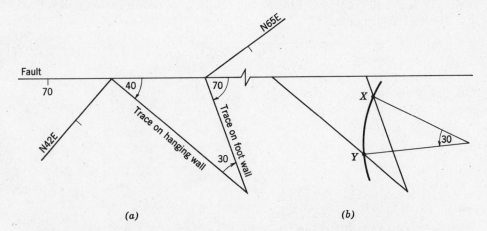

(a)   (b)

FIGURE 15.15 Angle of rotation. (a) Map view with superimposed construction of the traces of the fault plane of a single displaced plane. The angle of rotation is defined by the angle between these two traces. (b) The center of rotation constructed from displaced points X and Y.

EXERCISES

1. The plane of a normal-slip fault strikes due north and dips 60° to the west. The fault displaces a structural plane (N 90 W, 30 N) which shows 100 m of left-separation. What is the amount of the net slip?

2. A fault (N 90 W, 60 N) cuts two structural planes: Plane 1 = N 45 W, 30 NE;
   Plane 2 = N 50 E, 45 NW. The amounts and senses of separation are shown in Fig.
   X15.1. What is the amount and orientation of the net slip, what are the dip-slip
   and strike-slip components, and what is the name of the fault?

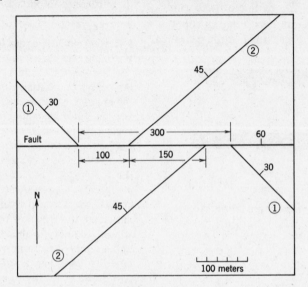

FIGURE X15.1

3. A fault (N 30 E, 60 W) displaces two planes as shown in Fig. X15.2. What is the
   angle and sense of rotation? Locate the center of rotation which will account for
   the observed displacements.

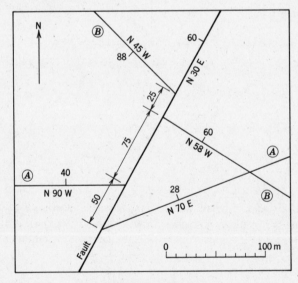

FIGURE   X15.2

16
Geometry of Stress

FORCE AND STRESS

Force is defined by Newton's Second Law as that which accelerates a mass. Force has both magnitude and direction and is therefore a vector quantity. Being vectors, forces can be treated according to the rules of vector algebra. Two manipulations are particularly important: (1) resolving a single force into components, and (2) finding the single resultant of several forces.

If two or more external forces are balanced in such a way that the mass neither accelerated nor rotates it is said to be in *equilibrium* (Fig. 16.1). This requires that the sums of the forces and moments are zero.

$$\Sigma\, F_x = \Sigma\, F_y = \Sigma\, F_z = 0$$
$$\Sigma\, M_x = \Sigma\, M_y = \Sigma\, M_z = 0$$

$$(16.1)$$

The action of the external forces will then be distributed internally. This action can be conceived of in terms of internal forces acting between various parts of the body mass. For example, in Fig. 16.1, the left half of the body can be considered to be pressing against the right half across the plane AB. The magnitude of this action is defined by its intensity, that is, by the amount of force per unit area of the surface on which it acts. This intensity is termed *stress*. In general, the stress will not be uniform over the area. Consequently, it is necessary to define stress more strictly as the limiting value of the ratio of force to area, as the area is contracted, or

$$S = \frac{\lim}{\Delta A \to 0}\ \frac{\Delta F}{\Delta A} \qquad (16.2)$$

The value of S on the small area in the vicinity of some point is referred to as the stress at that point (Fig. 16.2a). The direction of this stress is the limiting direction of the resultant of ΔF. Generally, it will be inclined to the area ΔA, and it is usual to resolve such a stress into two components (Fig. 16.2b): a normal stress (σ) perpendicular to the area, and a shearing stress (τ) acting in the plane of the area.

The stress will depend on both the nature of the system of applied forces and on the position and orientation of ΔA with respect to this system. The way the stress at point O varies as the orientation of ΔA varies is of special interest. To this end, it is convenient to note the components of the stress acting

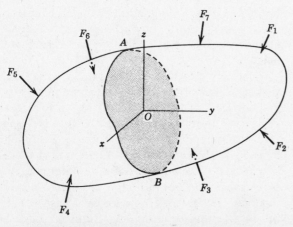

FIGURE 16.1 A body in equilibrium.

141

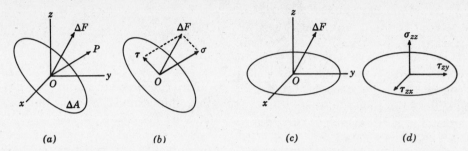

(a)   (b)   (c)   (d)

FIGURE 16.2 Stress at a point, and the components of stress acting on an element of area.

on the plane in the directions of the coordinate axes: generally these will be S_x, S_y and S_z. However, if ΔA is oriented so that it lies in one of the coordinate planes there will be shear stress components parallel to two of the coordinate axes and a normal stress component parallel to the third. For example, if ΔA lies in the xy plane (Fig. 16.2c,d), the components will be σ_{zz}, τ_{zy}, and τ_{zx}. The first suffix in this notation indicates the direction of the normal to ΔA, and the second the direction in which the stress component acts. There will be similar components when ΔA is parallel to the other two coordinate planes. In all, nine such stress components can be associated with point O (see Fig. 16.3).

$$\begin{bmatrix} \sigma_{xx} & \tau_{xy} & \tau_{xz} \\ \tau_{yx} & \sigma_{yy} & \tau_{yz} \\ \tau_{zx} & \tau_{zy} & \sigma_{zz} \end{bmatrix} \qquad (16.3)$$

The normal stress components can be, and often are simply written σ_x, σ_y, and σ_z, without loss of meaning. However, the retention of the double suffixes throughout preserves symmetry and anticipates a more general notation.

Finally, the signs of the stress components must be specified. Unfortunately there are a number of different sign conventions. The most widely accepted one in engineering and continuum mechanics is that a stress component is *positive* if it acts in a positive direction on a surface whose outward normal acts in a positive direction, or in a negative direction on a surface whose outward normal acts in a negative direction; otherwise it is *negative*. Thus the components shown in Fig.

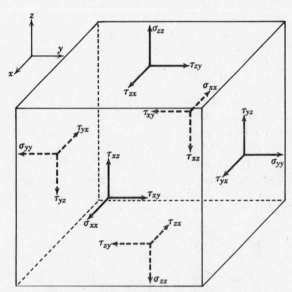

FIGURE 16.3 Stress components in three dimensions.

16.3 are all positive. In contrast, where compressive normal stresses are dominant, as in the fields of soil and rock mechanics and in structural geology, it is convenient to reckon them positive. This will be the convention used here, and accordingly, the stress components of Fig. 16.3 are all considered negative.

Although the units of stress and their actual values have no direct bearing on the geometrical aspects of stress, they are important in any application. In the Systéme International d'Unités (designated SI in all languages) the *newton* is defined as that force which gives to a mass of one kilogram an acceleration of one meter per second per second. The unit of pressure or stress is then the newton per square meter (N/m²). Because 1 N/m² is a very small pressure, it is convenient to use mega-newtons per square meter (1 MN/m² = 10^6 N/m²). Although not an SI unit, the bar (1 bar = 10^5 N/m²) and the kilobar (1 kb = 10^3 bar) are widely used, especially in experimental geology. It is also useful to have some idea of what constitutes geologically realistic magnitudes of stress. In average crustal rocks, the following easily remembered approximation relates the vertical stress and depth of burial in the absence of any tectonic stress: 1 m ≈ 0.25 bar or 1 km ≈ 0.25 kb.

STRESS IN TWO DIMENSIONS

Given the nine components, the stress on any inclined plane can be determined. In order to indicate the method, and also because it is important in its own right, some aspects of the geometry of stress in two dimensions will be explored. Further details can be found in Ramsay (1967, p. 27), Jaeger and Cook (1969, p. 7), Johnson (1970, p. 337), and in many books on elasticity (e.g. Timoshenko and Goodier, 1951, p. 14). Following conventional practice the xy plane will be used, and all quantities will be considered independent of z, which therefore will not enter the formulation. This restriction reduces the nine components to just four (see Fig. 16.4).

$$\begin{bmatrix} \sigma_{xx} & \tau_{xy} \\ \tau_{yx} & \sigma_{yy} \end{bmatrix} \qquad (16.4)$$

Further, the equilibrium condition requires that the moment about the z axis be zero. This implies that

$$\tau_{xy} = \tau_{yx} \qquad (16.5)$$

and this reduces the independent stress components to three in number.

Uniaxial stress. Given an element with thickness t subjected to a uniform compressive stress in the x direction, we wish to find the stress on any inclined plane AB defined by θ, the angle the normal makes with the x axis (Fig. 16.5a). To do so we imagine a triangular "free-body" cut from the mass (Fig. 16.5b). To restore this body to equilibrium we must replace the action of the originally adjacent material with equivalent forces. Remembering that stress is defined as force per unit area ($S = F/A$), then the forces can be determined from the stress ($F = S \cdot A$). If the inclined plane has an area $a = AB \cdot t$, then the area of side $OA = a \cdot \cos \theta$. The force on this side is then $\sigma_{xx} a \cos \theta$, and an equal and opposite force must also act on plane side AB. From the triangle of forces (Fig. 16.5c),

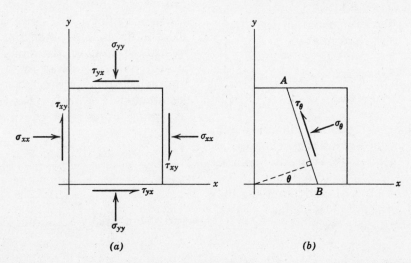

(a)   *(b)*

FIGURE 16.4 Stress in two dimensions. (*a*) Positive components (compare with the xy plane of Fig. 16.3). (*b*) Stress components acting on the inclined plane AB defined by θ, the angle the normal makes with the x axis.

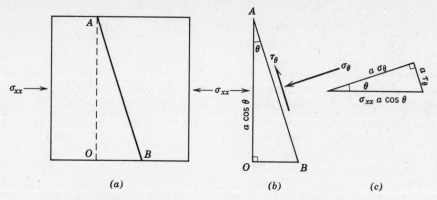

FIGURE 16.5 Uniaxial normal stress acting parallel to the x axis.

the following relationships hold involving the components normal and tangential to AB:

$$\cos \theta = a \cdot \sigma_\theta / (\sigma_{xx}\, a \cos \theta)$$
$$\sin \theta = a \cdot \tau_\theta / (\sigma_{xx}\, a \cos \theta)$$

Solving for σ_θ and τ_θ, the normal and shearing stresses on the inclined plane defined by θ, gives

$$\sigma_\theta = \sigma_{xx} \cos^2 \theta \qquad (16.6a)$$

$$\tau_\theta = \sigma_{xx} \sin \theta \cos \theta \qquad (16.6b)$$

Biaxial Stress. To determine the stress on AB when normal stresses act in both coordinate directions, we first consider the effect of the stress in the y direction. Proceeding just as before, the force in side AB is $\sigma_{yy} = a \sin \theta$ (Fig. 16.6b). Then from the triangle of forces (Fig. 16.6c), the components are

$$\sigma_\theta = \sigma_{yy} \sin^2 \theta \qquad (16.7a)$$

$$\tau_\theta = \sigma_{yy} \sin \theta \cos \theta \qquad (16.7b)$$

From the principle of superposition, the total stress on the inclined plane is the sum of the stresses due to the two separate uniaxial normal stresses. Adding (16.6a) and (16.7a), and (16.6b) and (16.7b), and observing that the contributions σ_{xx} and σ_{yy} make to the shearing stress are opposite:

$$\sigma_\theta = \sigma_{xx} \cos^2 \theta + \sigma_{yy} \sin^2 \theta \qquad (16.8a)$$

$$\tau_\theta = (\sigma_{xx} - \sigma_{yy}) \sin \theta \cos \theta \qquad (16.8b)$$

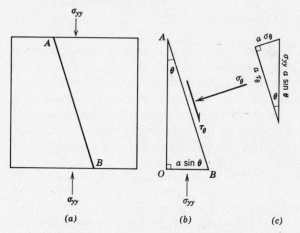

FIGURE 16.6 Uniaxial normal stress acting parallel to the y axis.

Pure Shear Stress. Finally, we seek the effect of the applied shear stresses alone (Fig. 16.7a). As before, the forces due to the applied stresses are determined. Considering the components related to each of the shear stresses separately, triangles of forces are constructed for each (Fig. 16.7c). From triangle of forces 1, the contribution by τ_{xy} is

$$\sigma_\theta = \tau_{xy} \sin \theta \cos \theta$$

$$\tau_\theta = \tau_{xy} \cos^2 \theta$$

And from triangle 2, the contribution by τ_{yx} is

$$\sigma_\theta = \tau_{yx} \sin \theta \cos \theta$$

$$\tau_\theta = \tau_{yx} \sin^2 \theta$$

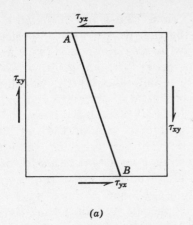

(a)   *(b)*   *(c)*

FIGURE 16.7 Pure shear.

Superimposing the separate contributions, paying attention to the sense of the shear stress components, then using (16.5) gives:

$$\sigma_\theta = 2\,\tau_{xy} \sin\theta \cos\theta \qquad (16.9a)$$

$$\tau_\theta = \tau_{xy}(\cos^2\theta - \sin^2\theta) \qquad (16.9b)$$

General Two Dimensional Stress. Now the contribution of all the components to the stress on the inclined plane can be expressed by adding (16.8a) and (16.9a), and (16.8b) and (16.9b) giving

$$\sigma_\theta = \sigma_{xx}\cos^2\theta + \sigma_{yy}\sin^2\theta$$
$$+ 2\,\tau_{xy}\sin\theta\cos\theta \qquad (16.10a)$$

$$\tau_\theta = (\sigma_{xx} - \sigma_{yy})\sin\theta\cos\theta$$
$$+ \tau_{xy}(\cos^2\theta - \sin^2\theta) \qquad (16.10b)$$

The variation of the normal stress with orientation is of special interest. The condition for maximum or minimum values of σ_θ is that $d\sigma_\theta/d\theta = 0$. Differentiating (16.10a) and setting equal to zero:

$$(\sigma_{yy} - \sigma_{xx})\sin\theta\cos\theta$$
$$+ \tau_{xy}(\cos^2\theta - \sin^2\theta) = 0 \qquad (16.11)$$

Substituting the trigonometric identities:

$$\sin\theta\cos\theta = \tan\theta/(1 + \tan^2\theta)$$

$$\cos^2\theta - \sin^2\theta = (1 - \tan^2\theta)/(1 + \tan^2\theta)$$

$$\tan 2\theta = 2\tan\theta/(1 - \tan^2\theta)$$

into (16.11), which then becomes

$$\tan 2\theta = 2\,\tau_{xy}/(\sigma_{xx} - \sigma_{yy}) \qquad (16.12)$$

This defines two angles, 2θ and $2\theta + 180°$; these in turn fix the orientation of two mutually perpendicular planes with the angles θ and $\theta + 90°$ on which the shear stress is zero. These planes are called the principal planes, and the normal stresses acting on them have the extreme values. These are the principal stresses.

If coordinate axes are chosen to coincide with the directions of these normal stresses on the principal planes, (16.10) simplifies to (16.8). Assuming that $\sigma_{xx} > \sigma_{yy}$, the maximum principal stress ($\sigma_1 = \sigma_{xx}$), and the minimum principal stress ($\sigma_3 = \sigma_{yy}$) are defined. Substituting the identities

$$\sin^2\theta = (1 - \cos 2\theta)/2$$

$$\cos^2\theta = (1 + \cos 2\theta)/2$$

$$\sin\theta\cos\theta = (\sin 2\theta)/2$$

and rearranging, (16.8) becomes

$$\sigma_\theta = \frac{\sigma_1 + \sigma_3}{2} + \frac{\sigma_1 - \sigma_3}{2}\cos 2\theta \qquad (16.13a)$$

$$\tau_\theta = \frac{\sigma_1 - \sigma_3}{2}\sin 2\theta \qquad (16.13b)$$

Before describing methods of solving these equations, the conditions under which stresses may be considered two dimensional must be made clear. One is when all the components involving z are zero ($\sigma_{zz} = \tau_{xz} = \tau_{yz} = \tau_{zx} = \tau_{zy} = 0$). This is the state of plane stress, but is strictly valid only for very thin plates. In all other general situations there are *three* mutually perpendicular principal stresses at each point. As it turns out, the value of the intermediate principal stress (σ_2) generally plays only a secondary role in the deformation and failure of material, and therefore, it is the $\sigma_1 \sigma_3$ plane which is usually of greatest interest. However, if the relationships between the stress components in three dimensions were of interest, with appropriate changes in the principal stresses involved, the two dimensional equations (16.13) are valid for stress components acting on any two of the three principal planes.

Mohr's Circle for Stress. The final equations for two dimensional stress (16.13) are identical in form with the equations for finite strain (6.8), and as with strain are most easily solved graphically with a Mohr's circle (Fig. 16.8). This construction consists of a circle with a radius of $(\sigma_1 - \sigma_3)/2$ units, centered on the horizontal axis $(\sigma_1 + \sigma_3)/2$ units from the

origin. The values of the normal stress defined by the two points of intersection of the circle with this horizontal axis are the principal stresses. The angle 2θ locates a point on the circumference with coordinates (σ_θ, τ_θ), which are the stress components on the inclined plane defined by θ. In order to solve problems, one of the previously established conventions must be modified. If a shear stress component tends to rotate the element in an anticlockwise direction it is reckoned positive, and the associated point is plotted above the horizontal σ axis. Also, if θ is measured anticlockwise from the reference axis it is considered positive.

There are two general types of problems that can be solved with this method: (1) from known principal stresses, the stress components on any inclined plane can be determined, and (2) from a general state of stress, the magnitudes and directions of the principal stresses can be found.

PROBLEM 1

From known values of the principal stresses ($\sigma_1 = +10$, $\sigma_3 = +2$), find the stress components on the inclined plane whose normal makes an angle of $\theta = 45°$ with the σ_1 axis (Fig. 16.9a).

METHOD (Fig. 16.9b)

1. In a two dimensional cartesian system, plot points representing the principal stresses $\sigma_1 = 10$

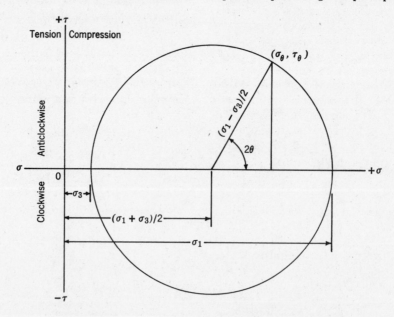

FIGURE 16.8 Mohr's circle for stress in two dimensions.

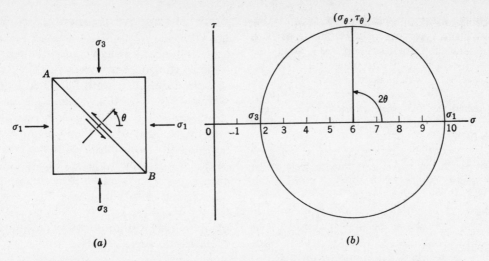

FIGURE 16.9 Stress components on any inclined plane given the principal stresses.

units and σ_3 = 2 units along the horizontal axis, measuring from the origin using a convenient scale.
2. With a point $(\sigma_1 + \sigma_3)/2$ = 6 units from the origin as the center, draw a circle passing through the points representing the two principal stresses.
3. From the center of this circle locate a line of radius with an angular distance 2θ from σ_1. In this example, the plane of interest is defined by θ = $+45°$, and therefore 2θ = $+90°$ is plotted (measured anticlockwise).

ANSWER

The coordinates of the point on the circumference gives the values of σ_θ = 6, and τ_θ = 4. The normal stress on this plane is equal to $(\sigma_1 + \sigma_3)/2$, which is also the mean stress in two dimensions. The shear stress on this plane has the maximum possible value of $(\sigma_1 - \sigma_3)/2$. Note that the sense of shear is positive when the angle θ is positive (this sense can also be obtained by inspection from Fig. 16.9a).

PROBLEM 2

From the three independent stress components (σ_{xx} = +10, σ_{yy} = +4, and τ_{xy} = +3), find the principal stresses and their orientation (Fig. 16.10a).

METHOD (Fig. 16.10b)

1. Plot the values of σ_{xx} = 10 units and σ_{yy} = 4 units along the horizontal σ-axis.
2. The shear stress components associated with each normal stress are plotted parallel to the τ-axis. In the example τ_{xy} = +3 is plotted above the σ-axis, and τ_{yx} = -3 below.
3. A line connecting the points (σ_{xx}, τ_{xy}) and (σ_{yy}, τ_{yx}) locates the center of the circle and its diameter, which is then drawn.

ANSWER

The intercepts along the σ-axis give the values of σ_1 = 11.23 and σ_3 = 2.77. The angle 2θ = -45° between the x direction and the σ_1 axis established

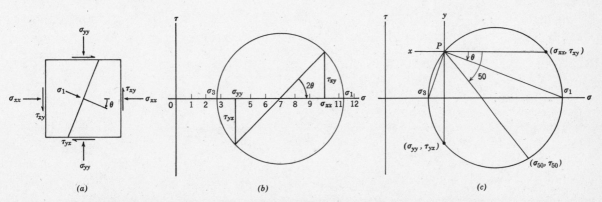

FIGURE 16.10 Principal stresses from general components in two dimensions.

the orientation of σ_1, which is $\theta = 22.5°$ in the physical plane, measured in the same sense. This construction illustrates an important additional point. The center of the circle is fixed by $(\sigma_{xx} + \sigma_{yy})/2$, which then must be equal to $(\sigma_1 + \sigma_3)/2$. The constant value of the sum $(\sigma_1 + \sigma_3)$ is called the *first stress invariant* (I_1), and can be used in the actual construction and in checking the results.

A slight modification of this construction can be used to give the same answer in a somewhat simpler, more direct manner. Point (σ_{xx}, τ_{xy}) represents the stress components acting on the plane perpendicular to the x-axis, and point (σ_{yy}, τ_{yx}) those components on the plane perpendicular to the

y-axis. If these two axes are transferred from the physical plane to the Mohr diagram so that they pass through the points representing their corresponding stresses, they will intersect the circle at a point P, called the *origin of planes* (Fig. 16.10c). From this origin, the stresses on any other plane can be determined. In Fig. 16.10c, the angle $\theta = 22.5°$, defining the orientation of σ_1, appears directly. Once this new origin is located, the stress components on any other plane can be determined with equal ease; for example, the figure illustrates how to find the stresses on a plane whose normal makes an angle of $50°$ with the x-axis.

EXERCISES

1. Derive equations (16.13a,b) directly from Mohr's circle (Fig. 16.8).

2. If $\sigma_1 = 2.0$ kb and $\sigma_3 = 0.5$ kb, what is the stress acting on a plane whose normal makes an angle of $20°$ with the σ_1 direction?

3. With a Mohr's circle construction, and the stress components given in Fig. X16.1:
   (a) Determine the magnitude and orientation of the principal stresses. Note that the units of stress are not specified.
   (b) Superimpose the stress systems of diagrams *b* and *c* and determine the magnitude and orientation of the principal stresses.
   (c) Rotate diagram *b* $30°$ clockwise, and then superimpose this new system with that of diagram *c* and determine the principal stresses. Note that stress components can be added only if they act on parallel planes.

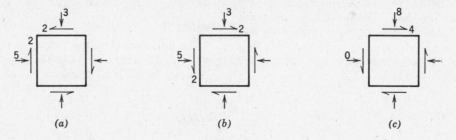

(a)   (b)   (c)

FIGURE   X16-1

17
Faults and Stresses

INTRODUCTION

For every general state of stress the mean stress $\bar{\sigma}$ can be defined.

$$\bar{\sigma} = (\sigma_1 + \sigma_2 + \sigma_3)/3 = I_1/3 \qquad (17.1)$$

In two dimensions, $\bar{\sigma}$ is represented by the center point of Mohr's circle. This portion of the total stress causes only a volume change. The degree to which the stress system departs from this mean is termed the deviatoric stress. The deviatoric principal stresses are

$$\begin{aligned} \sigma'_1 &= (\sigma_1 - \bar{\sigma}) \\ \sigma'_2 &= (\sigma_2 - \bar{\sigma}) \\ \sigma'_3 &= (\sigma_3 - \bar{\sigma}) \end{aligned} \qquad (17.2)$$

In two dimensions, the radius of Mohr's circle is a measure of the deviatoric stress. This part of the total stress causes distortions.

As a starting point in the analysis of what happens if a rock mass is subjected to tectonic stresses, it is convenient to assume that in the absence of such stresses, past and present, the stress at a point at depth is nondeviatoric, that is $P = \sigma_1 = \sigma_2 = \sigma_3$, and its value depends on the overburden:

$$P = \rho\, g\, z \qquad (17.3)$$

where ρ is the mean density, g the acceleration due to gravity, and z the depth of cover. In this hypothetical situation, the mean stress *is* the total stress. Anderson (1951, p. 13)

termed this the *standard state*. With the onset of the tectonic stress the state of stress at the point in question becomes deviatoric. The easiest way of illustrating this sequence of changes is with the aid of Mohr's circles. The standard state pressure can be represented by a point on the σ-axis. As the tectonic stress is superimposed on this state, the point becomes a circle which then grows in size (Fig. 17.1).

Geometrically, there is no limit to the size of the circle that might be drawn or conceived. Physically, however, there is a definite limit, for if the deviatoric stress becomes too large, the material will fail. To learn the details of

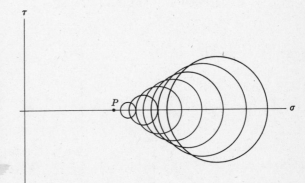

FIGURE 17.1 The changing state of stress at a point illustrated by a series of Mohr's circles.

149

rock behavior under various stress conditions it is necessary to resort to experiment. The ideal procedure is to apply to a rock specimen a stress system in which the three principal stresses can be systematically varied. Commonly, the tests fall short of this ideal. The method of the uniaxial comprehensive test is to apply pressure to the ends of rock cylinder by means of a press; the specimen is otherwise exposed to the air so that $\sigma_2 = \sigma_3 = 0$. In the conventional triaxial test the cylinder is immersed in a fluid under hydrostatic pressure ($\sigma_2 = \sigma_3 =$ confining pressure). In order to perform a true triaxial test, where $\sigma_1 \rangle \sigma_2 \rangle \sigma_3$, a rectangular specimen is compressed between two sets of mutually perpendicular anvils with σ_3 being applied through a fluid confining medium. This requires an elaborate device, and only a limited number of such tests have been performed (see Mogi, 1971).

Briefly the results of such tests show that the conditions required to produce failure are related to the deviatoric stress, but also that the higher the mean stress, the higher the required deviatoric stress. Further, they show that whether the failure occurs by fracture or flow is also stress dependent (Fig. 17.2). Two types of fractures may occur: extension fractures parallel to the σ_1 σ_2 plane (Fig. 17.2a), or shear fractures related to but not identical with the planes of maximum shearing stress. If fracture occurs before negligible permanent strain, the material is said to be *brittle* (Fig. 17.2b); if a small plastic strain ($< 5\%$) preceeds the fracturing it is *semibrittle* (Fig. 17.2c). Experimentally produced shear frac-

tures often occur in conjugate pairs. The intersection of these planes is parallel to σ_2, and the acute angle between them is bisected by σ_1.

What conditions have to be met before a rock will fail by shear fracture? Consider first the case where a fracture (that is, a surface of no cohesion) already exists in the rock mass. Renewed sliding will take place when the resolved shear stress on the plane is sufficient to overcome the friction and the effect of the normal stress across the plane. This situation is analogous to the familiar problem of a block sliding on an inclined plane (Fig. 17.3a). The force acting on such a block of mass m is

$$F = mg \qquad (17.4)$$

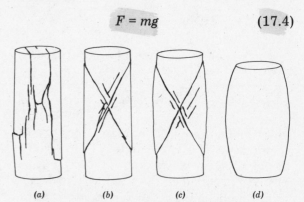

FIGURE 17.2 Types of failure illustrated by experimentally deformed limestone at varying confining pressures (see Verhoogen, et al., 1970, p. 458): (a) extension fractures (1 bar), (b) brittle shear failures (35 bars), (c) semibrittle shear failure (300 bars), (d) plastic failure (1000 bars).

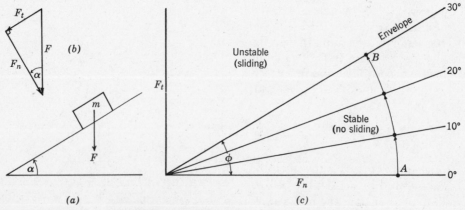

FIGURE 17.3 Sliding block on an inclined plane.

This force can be resolved into two components, one tangential to the inclined surface, and the other normal to it (Fig. 17.3b). The following relationships then hold

$$\cos a = F_n/F \text{ or } F_n = mg \cos a$$
$$\sin a = F_t/F \text{ or } F_t = mg \sin a \qquad (17.5)$$

It is instructive to plot the values of F_n and F_t as a function of the angle of inclination. As the angle increases, the point representing the force components describes a curve, such as AB in Fig. 17.3c. At some value of the angle of inclination the frictional resistance of the two contacting surfaces will be overcome, and the mass will slide. This limiting angle is termed the angle of friction ϕ.

It is also convenient to define a coefficient of friction μ.

$$\mu = F_t/F_n = \tan \phi \qquad (17.6)$$

On the graph, the line with slope equal to tan ϕ is the boundary, or envelope separating those ratios of F_t/F_n which are stable and those which are unstable.

Now consider the element of rock containing a plane of no cohesion at some moderate depth. The original standard state pressure is given by (17.3), and the plane will have associated with it some coefficient of sliding friction. The progressive change in the state of stress is represented by a series of Mohr's circles (Fig. 17.4). Sliding will take place when the ratio of the shearing to normal stress on the plane is equal to the coefficient of friction.

$$\tau/\sigma = \tan \phi \qquad (17.7)$$

Thus the first possibility of renewed faulting occurs when the circle is tangent to the envelope (Fig. 17.4c). However, it will occur then only on a plane whose *normal* measured from the σ_1 direction is given by

$$2\theta = 90 + \phi \quad \text{or} \quad \theta = 45 + \phi/2 \qquad (17.8)$$

Or alternatively, on the *plane* making an angle a with the σ_1 direction, where a is given by

$$a = 45 - \phi/2 \qquad (17.9)$$

For a plane of some other orientation, a further increase in the deviatoric stress is required to produce sliding (Fig. 17.4d,e).

But what of material without such pre-existing planes? Two hundred years ago, Coulomb (see Handin, 1969) suggested that the shear stress which tends to cause failure is resisted in two ways: (1) by the normal stress across the potential shear plane, and (2) by the cohesive shear strength of the material. This is now known as the Coulomb criterion of shear failure, and is usually written

$$\tau = \tau_0 + \sigma \tan \phi \quad \text{or} \quad \tau = \tau_0 + \mu\sigma \quad (17.10)$$

where τ_0 is the cohesive strength or shear strength at zero normal stress, ϕ is, by analogy with sliding friction, the angle of internal friction, and μ is the coefficient of internal

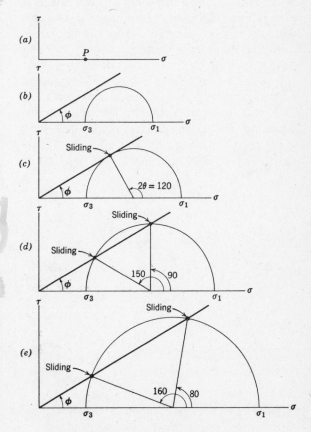

FIGURE 17.4 Conditions required for renewed sliding on variously oriented planes of no cohesion, assuming an angle of friction ϕ = 30°.

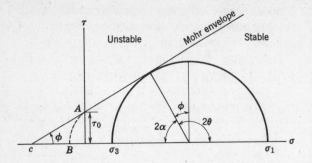

FIGURE 17.5 The Mohr envelope representing the Coulomb criterion of failure. The Mohr's circle illustrated the stresses of failure.

friction. The cohesive strength is of the order of 100-200 bars for sedimentary rocks and 500 bars for crystalline rocks. The average value of the angle of friction is about 30°.

The Coulomb criterion of shear failure can be expressed graphically on a Mohr diagram (Fig. 17.5). The only difference between this and the previous diagram is that the boundary between stable and unstable stresses, usually called the Mohr envelope, does not pass through the coordinate origin. Compressive tests made at low to moderate confining pressures confirm that the linear envelope describes the shear failure conditions to a good approximation. On the other hand, the predicted tensile strength ($= \tau_0 \cot \phi$) is too

great; commonly it is only about half the cohesive strength. The envelope can be empirically modified by the dashed curve AB in Fig. 17.5 to take this into account. The linear relationship also implies an ever increasing strength at higher values of $\bar{\sigma}$. Not only is this intuitively suspect, but experiments show that the curve finally becomes concave downward. More elaborate criteria have been proposed to take these and other factors into account (see Price, 1966, p. 29f). However, because of its simplicity the Coulomb criterion is still the most important for many purposes.

The analogy between the Coulomb criterion and the conditions for sliding is useful both as a memory aid, and as a simple physical explanation. There is however a fundamental flaw—what is internal friction? In loose, dry sand it is of the nature of external or sliding friction, but for cohesive material it can neither be measured directly nor is there an adequate physical explanation. It should probably be regarded as no more than the slope of the Mohr envelope.

CLASSIFICATION OF FAULTS

Clearly, no shearing stress acts along the air-earth interface and one of the principal stresses must therefore be perpendicular to

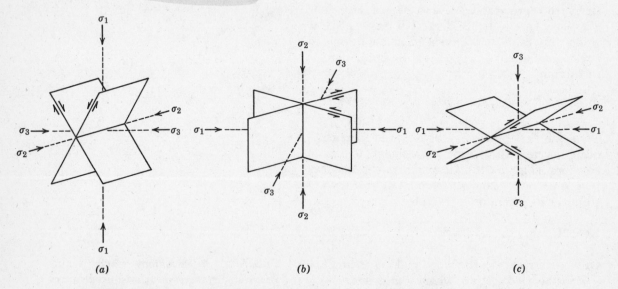

FIGURE 17.6 Dynamic classification of faults. (a) normal faults (σ_1 vertical), (b) wrench faults (σ_2 vertical), (c) thrusts (σ_3 vertical).

the earth's surface. In many areas this principal stress can be considered to be approximately vertical. In combination with the geometrical relationship between fracture planes and stress directions, this leads to a dynamic classification of faults.

1. If σ_1 is vertical, the fault dip at an angle of $a = 45 + \phi/2$. These are *normal faults* (Fig. 17.6a).

2. If σ_2 is vertical, the faults planes are also vertical, and the slip direction is horizontal. These are *wrench faults* (Fig. 17.6b).

3. If σ_3 is vertical, the fault planes dip at an angle $a = 45 - \phi/2$; these are *thrusts* (Fig. 17.6c). Note the genetic connotation here.

FAULTS AND STRESSES

It is now possible to solve certain problems dealing with the geometric relationship between faults and stresses. The pertinent features are:

1. The intersection of a pair of conjugate faults defines the orientation of σ_2.

2. The acute angle between conjugate pairs is $2a$, and is bisected by σ_1.

3. The slip direction is defined by the intersection of the fault plane and the $\sigma_1\sigma_3$ plane.

4. The sense of slip is such that the wedge of material in the acute angle moves inward along σ_1 (see Fig. 17.6).

PROBLEM

Given two faults with attitudes N 24 W, 50 W, and N 48 W, 76 NE, and assuming they are conjugate, find the orientation of the principal stresses, the direction and sense of slip and the angle of internal friction.

METHOD (Fig. 17.7)

1. Plot the faults as great circle on the stereonet. The point of intersection defines σ_2.

2. Draw in the great circle for which σ_2 is the pole. This is the $\sigma_1\sigma_3$ plane, and its intersection with the faults fixes the slip directions.

3. Bisect the acute segment of the $\sigma_1\sigma_3$ great circle between the faults to locate σ_1; σ_3 is 90° along this same great circle. The directions of τ_{max} are 45° from σ_1 (and σ_3).

4. The angle between σ_1 and one of the slip directions is the angle a; the angle of internal friction $\phi = 90 - 2a$.

ANSWER

The orientations of the principal stresses are: σ_1 (68, N 55 E), σ_2 (21, N 43 W), and σ_3 (13, N 54 E). The angle of internal friction is 32°. Since σ_1 is close to vertical, the displacement on the

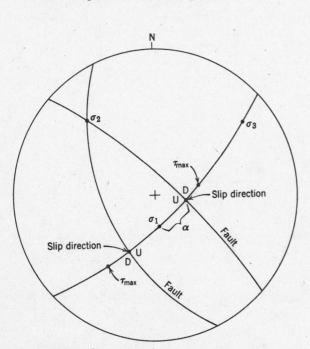

FIGURE 17.7 Stress directions determined from conjugate faults.

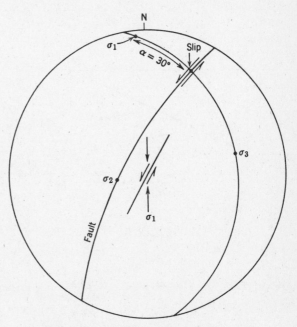

FIGURE 17.8 Stress directions estimated from a single fault with known slip directions.

faults must be dominantly normal slip. This sense can also be obtained by visualizing each fault (with the flattened hand) and σ_1 direction (with the index finger).

If less information is available then, of course, fewer conclusions concerning the full relation between the fault and the stress directions can be obtained. However, an approximate answer may still be possible.

PROBLEM

Given a dominantly left-slip fault with an attitude of N 30 E, 70 W, and a slip direction which pitches 15 N, locate the principal stress directions as accurately as possible.

APPROACH

The σ_2 direction and the $\sigma_1 \sigma_3$ plane can be obtained by previous methods. An estimate of the value of ϕ then fixes σ_1. Without additional information use of the average value of $\phi = 30°$ is probably the best one can do.

METHOD (Fig. 17.8)

1. Plot the great circle representing the fault, and the slip direction as a point on this trace. The σ_2 direction is found by counting off 90° from the slip line. Since σ_2 is the pole of the $\sigma_1 \sigma_3$ plane, its great circle is easily added.

2. With the estimated value of ϕ, calculate $a = 45 - \phi/2 = 30°$ (note: when $\phi = 30$, $a = 30$, a special case). Locate the σ_1 direction 30° from slip direction along the $\sigma_1 \sigma_3$ great circle. It is critical that σ_1 be plotted on the correct side of the slip line, and this depends on sense of slip. In order to do this, it is necessary to visualize the orientation of σ_1 compatible with the known slip-sense, and again, the use of the hands is a great aid. It may also help to rotate the fault plane, either literally or visually into some other position. Here, for example, it is easy to imagine the fault rotated into a vertical orientation with the slip direction horizontal. In this position, the σ_1 direction would have to be north-south (see inset, Fig. 17.8), and it is then clear that its representation must be plotted on the north side of the fault trace.

ANSWER

The orientation of the principal stresses are σ_1 (2, N 3 W), σ_2 (65, S 83 W), and σ_3 (25, N 88 E). Other estimates will, of course, give other orientations of σ_1 and σ_3, though the probability of great error is small.

Fractures, often filled with quartz or calcite, are associated with some faults. These are extension fractures (cf. Fig. 17.2a), and their presence give supplemental information about the stress and the displacement.

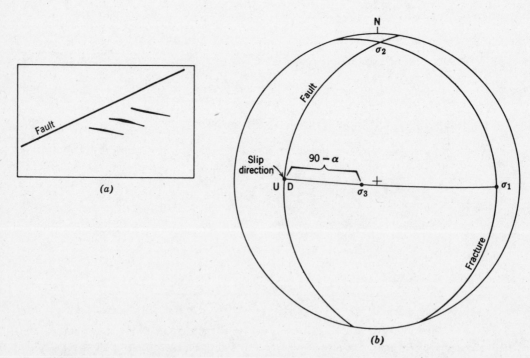

FIGURE 17.9 Stress directions from a fault and associated extension fractures.

PROBLEM

A fault has an attitude of N 10 E, 25 W. Sub-horizontal quartz veins (N 20 W, 10 E) are present adjacent to the fault plane (Fig. 17.9a). Determine the orientation of the principal stresses, the angle of internal friction, and the direction and sense of slip.

METHOD (Fig. 17.9b)

1. Plot the plane of the fracture as a great circle; this is the $\sigma_1 \sigma_2$ plane.
2. Also plot the fault plane as a great circle. Its intersection with the $\sigma_1 \sigma_2$ plane locates σ_2.
3. Using σ_2 as the pole, the $\sigma_1 \sigma_3$ plane can be drawn. This great circle intersects the fault at the slip direction, and the fracture trace at σ_1.
4. The σ_3 direction is 90° from σ_1. The angle between σ_1 and the slip direction is a, or, as here, measuring from σ_3, the angle is 90 - a. The angle of internal friction $\phi = 90 - 2a$.

ANSWER

The orientations of the principal stresses are σ_1 (10, S 87 E), σ_2 (4, N 2 E), and σ_3 (80, S 71 W). The angle of internal friction is 20°. The sense of slip is reverse, and the fault is a thrust.

FAULTS IN ANISOTROPIC ROCKS

If a rock mass is anisotropic the geometrical relationship between the principal stress directions and the faulting is more complicated. Two types of structures may influence faulting in this way: pervasive planar fabrics, such as cleavage and planes of discontinuity with low or zero cohesive strength, such as joints.

Donath (1961, 1964) experimentally investigated the role of cleavage in faulting. By loading to failure a series of cylinders of slate cut at different angles, a varied relationship between the attitude of the shear fracture plane and the cleavage was found (see Fig. 17.10). If an isotropic rock had been used in these experiments a conjugate pair of fractures inclined to σ_1 at an angle of approximately 30° would have been expected. In the slate such fractures were obtained in only two situations: in one experiment at higher confining pressure with the cleavage parallel to

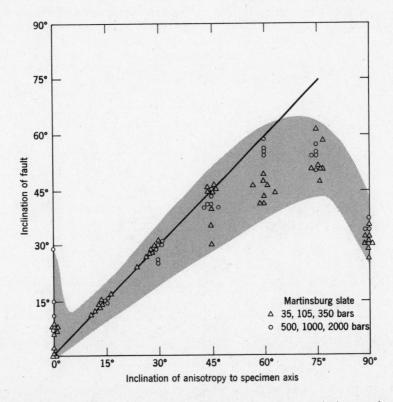

FIGURE 17.10 Relationship between fracture orientation and planar anisotropism. (After Donath, 1963.)

the long axis of the cylinder and σ_1, and in all experiments with the cleavage perpendicular to σ_1. In these two configurations the slate is effectively isotropic. In all others only one fracture developed, and its attitude was controlled, either directly or indirectly, by the cleavage. This is most clearly demonstrated by the essentially 1:1 relationship between cleavage and fracture attitude on the left side of the graph. Although for larger angles of inclination, the fractures no longer parallel the cleavage, angles greater than 45° indicates a definite influence by the planar weakness. Jaeger (1969, p. 170) presented a generalized version of the Coulomb criterion to include the effect of such anisotropism which is in reasonable agreement with these experiments.

From the point of view of interpreting field examples, however, the uncertainties introduced by effect of the anisotropism makes the determination of the principal stress directions impossible. This can be illustrated more clearly by considering the second type of structure—a single plane of weakness. In three dimensions the direction of the resolved shearing stress on a plane inclined to the principal stress directions depends on the magnitude of the principal stresses, and is not, in general, simply related to the principal directions. It follows, then, that if this shear stress exceeds the resistance on the plane, the direction of slip will also not be related to the principal stress directions. Therefore, the orientation of the stress field responsible for renewed movement on a single fault is also unsolvable from geometric considerations alone.

There is one situation where stress orientation can be estimated from a knowledge of renewed movement on preexisting planes. If a rock mass is cut by abundant, variously oriented fractures, renewed slip will occur on those planes with approximately the orientation of the conjugate pair that would have formed in intact rock. Given such fractures, the analytical procedure consists simply of plotting the poles of the planes showing evidence of such movement on a stereogram. Barring a preferred orientation of the preexisting plane, the points will form two clusters which will bear a symmetrical relationship to the principal stress directions (Fig. 17.11). Compton (1966, p. 1370) has applied a similar technique with interesting results.

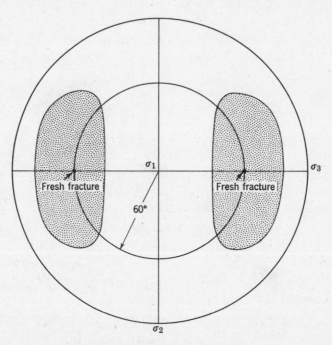

FIGURE 17.11 Assuming an average angle of friction, preexisting fracture planes with poles falling in the shaded zones will show renewed movement. (After Jaeger, 1969, p. 161.)

LIMITATIONS

These relationships between stresses and faulting are fundamental: if the shear strength of a rock mass is exceeded it will fracture, or if the resistance on a plane is surpassed it will slip. However, the interpretation of real, large-scale faults is fraught with additional difficulties. For example, regionally extensive wrench faults may well extend depths beyond the zone of semibrittle fracture. The slip then may well be related to deep crustal or subcrustal flow, and for such flow the relatively simple stress-strain relation on which the fracture analysis is based does not hold. Deep flow would, of course, set up stresses in the overlying brittle rocks, but these would not necessarily be uniform along the entire length of the fault. Such nonuniformity, as interpreted from local fracture and fold analysis, has been demonstrated along parts of the San Andreas fault of California (Dickinson, 1966). The picture that emerges for this fault is one of series of irregular blocks or "tectonic rafts" along and adjacent to the fault. The stresses within adjacent rafts are at least partly local. This nonuniformity, together with the long history of movement, makes it difficult to interpret such faults in terms of a regionally developed stress field. Probably all such faults have a more complex origin and history.

EXERCISES

1. A rock has the following physical properties: cohesive strength τ_0 = 0.45 kb; angle of internal friction $\phi = 33°$.

    a. Assuming that during the build up of a tectonic stress field σ_3 remains a constant 1.0 kb, what will be the value of σ_1 at the point of shear failure?
    b. Under the same conditions a preexisting plane of no cohesion with an angle of sliding friction also of 33° is inclined 45° to the σ_1 direction, which will occur first, slip or fracture?
    c. Which will occur first if the preexisting plane is inclined 55° to the σ_1 direction?

2. Two faults have attitudes of (N 10 W, 60 E), (N 30 E, 70 W). What are the directions of the principal stresses, the angle of internal friction, and the sense and direction of slip? (Ans. σ_1 = 54, S 4 W).

3. The traces of two conjugate sets of shear planes are exposed in the walls and floor of a quarry. Determine as many of the aspects of the fault-stress relationship as possible.
    Floor (horizontal): traces trend N 25 W, and N 82 W.
    Face 1 (N 10 E, 70 E): traces pitch 48 N, and 18 S.
    Face 2 (N 76 E, 80 N): traces pitch 26 E, and 30 W.

4. Vertical extension fractures are associated with a fault (N 63 E, 70 S). Solve for the various factors describing the fault-stress relationship.

5. A fault (N 32 E, 60 NW) with slickensides (54, N 23 W) shows normal dip separation. Estimate the orientation of the principal stresses.

6. Of a large number of fractures, the following orientations show evidence of renewed slip. Estimate the orientation of the responsible stress field, and for fracture a estimate the direction and sense of slip.

(a)	N 30 E, 50 E	(g)	N 84 E, 16 S
(b)	N 5 W, 15 W	(h)	N 33 E, 63 SE
(c)	N 30 W, 30 W	(i)	N 4 E, 43 E
(d)	N 54 E, 66 SE	(j)	N 80 E, 20 N
(e)	N 50 E, 49 SE	(k)	N 12 E, 30 W
(f)	N 34 E, 28 NW	(l)	N 14 E, 56 E

18
Structure Contours

A structure contour is a line of equal elevation drawn on a structural surface; such contours are used to depict the form of the surface. In the simplest case, a structure contour drawn on a planar surface is a straight line and is synonymous with line of strike. On curviplanar surfaces, they appear as curved lines which are everywhere tangent to the strike. As an imaginary line connecting points of equal elevation structure contours are partially analogous to topographic contours. The visualization of the features portrayed by both types of contours follows the same rules. In addition, however, structure contours have several unique properties: the surface represented by the contours may overhang, or it may be broken by faults. In certain circumstances it may also be useful to use an inclined rather than horizontal datum plane (Conolly, 1936).

Structure contours are particularly important in showing the structures where the dip angles are small, and for showing the structure of large areas, for example, see the tectonic map of North America (King, 1969).

CONTOURING

The available factual information concerning the configuration of a structural surface usually takes the form of a series of points of known elevation. The construction of a structure contour map from this raw data involves two general steps.

Objective Interpolation. The choice of the contour interval depends on the amount of relief present on the surface, the map scale, and on the spacing of the data points and the accuracy of their location and elevation. Intermediate elevation points are then established which correspond to this chosen interval, and tentative contours drawn through them. The location of these intermediate points may be found by eye, especially if only a very few contours must be interpreted, with special devices (Marsh, 1960; Schweinfurth, 1969), or by simple construction.

PROCEDURE

1. Connect three adjacent elevation points with straight lines (for example, points A, B, and C of Fig. 18.1a, which are taken from the map of Fig. 18.2a).
2. By assuming that these three point define a planar portion of the surface, the location of the intermediate points can be found easily (Fig. 18.1b).

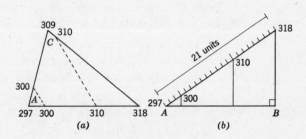

Fig. 18.1 Interpolation of intermediate elevation points.

158

a. Draw a line perpendicular to AB, through point B; this established two sides of a right triangle.

b. The elevation difference between points A and B is 21 m (= 318 - 297). Using any convenient scale, draw the hypotenuse of this triangle 21 units long.

c. Along this hypotenuse count off the vertical distances to even multiples of the contour interval. Here the interval is 10 m, so that points 3 and 13 units along this line represents 300 and 310 m contours. The location of these points are found by dropping perpendiculars back to line AB.

d. Repeat for the two other sides of the triangle ABC, and draw in the tentative contours as straight lines (Fig. 18.1a).

These intermediate points and tentative contour segments are only a first approximation of the form of the structural surface. For a curviplanar surface, curvilinear contours must, of course, be used. The next step is to draw these lines. An examination of the map may reveal advantageous places to start drawing, for example, in areas where close spacing of data points gives greater control, or where the steep slopes require a number of contours to be interpolated, or in areas of the highest or lowest elevations. Working in bands, curved lines are drawn which agree with the known elevations. In the absence of other information, the contours should be as smooth as possible, and with a spacing which is as nearly equal as possible. In general this will require that the contours do not pass exactly through the interpolated elevations, but this is understandable since the assumption which determined their location is false.

The contour patterns should progressively evolve during the work. It will be found that altering the position of one contour line requires that several adjacent ones must also be shifted. The final pattern of this stage of objective contouring is the one which technically accounts for the known elevations, but

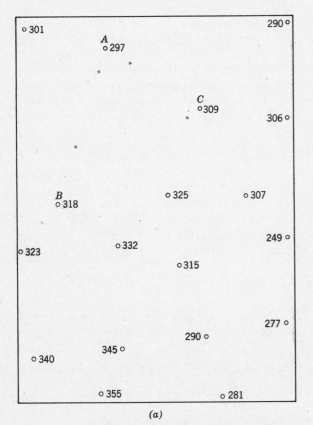

(a)

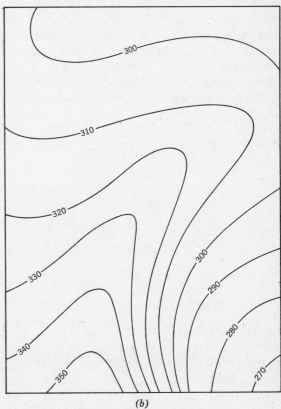

(b)

FIGURE 18.2 Method of contouring. (a) A map showing the elevation data. (b) The contoured equivalent.

introduces no features not demanded by them. The map will be easier to read if certain contours are shown with a heavy line, such as every fifth one, and if some or perhaps all are labeled with their value (Fig. 18.2b).

Contour License. The structure contour map resulting from this process can still only be an approximation of the true structural picture. It can always be improved upon by the skill and imagination of the geologist, and, of course, by adding more data points. The latitude in drawing the contours that allows imaginative expression is termed *contouring license*. Because the data reveal only a small part of the whole picture, there is always a question of how to represent the surface between these known points, especially when the spacing is large. The ability to make an intelligent interpretation in such case is part of becoming a capable geologist.

When objectively contouring, certain features may become evident that suggest several alternatives to the data. Fig. 18.3a shows the result of objective contouring; the steep, irregular zone in the central part of the map seems out of place. By a more careful contouring and by paying attention to the maintenance of nearly equal contour spacing and gradual changes in strike, the presence of

a fault suggests itself (Fig. 18.3b). In the second drawing the contouring should start on both sides of the suspect area and proceed independently toward it.

The wider the spacing of the control points, the greater the latitude for interpretation. Fig. 18.4a shows a routine projection of contours into a covered area for which no data is available. Fig. 18.4b shows quite a different structure in this same area: the same rhythmic folds that occur in the known area are continued into the covered portion. There is no direct evidence for the extra fold, but imagine its importance if it were an oil trap.

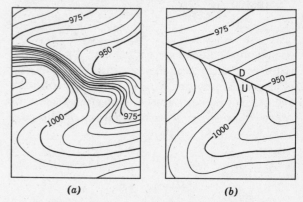

(a)   (b)

FIGURE 18.3 An example of contouring license. *(a)* An objectively contoured map. *(b)* An alternative interpretation involving a fault. (After LeRoy and Low, 1954, p.23)

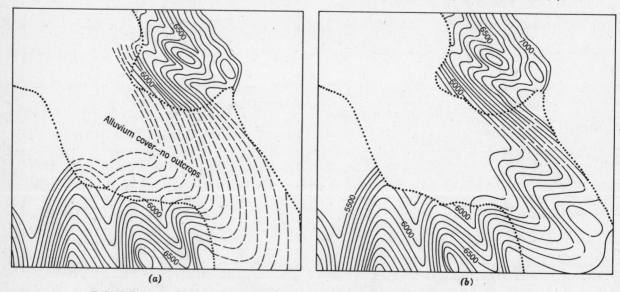

(a)   (b)

FIGURE 18.4 Projection of data into an unknown area. *(a)* Routine extrapolation. *(b)* Projection based on rhythmic character of the known folds. (From Low, 1957, Geologic Field Methods, Harper & Row. Used by permission.)

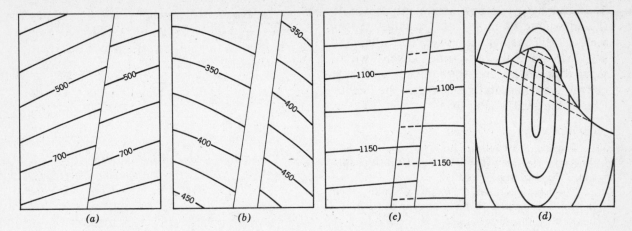

FIGURE 18.5 Structure contours on surfaces disrupted by faults: (a) vertical fault, (b) normal fault, (c) reverse fault, (d) faulted fold with contours on fault plane.

Structure Contours and Faults. Unlike the topographic surface, faults separate structural surfaces into two or more parts; this discontinuity of the surface also causes a discontinuity of the contours. Three different situations may arise.

1. Vertical faults appear to offset the contour lines (Fig. 18.5a).
2. Normal faults produce a gap between the lines (Fig. 18.5b).
3. Reverse or thrust faults causes an overlap of contours; the hidden contours are often shown by dashed lines (Fig. 18.5c).

The fault surface itself, especially if it has a curved shape, or is extensively developed, may also be represented by structure contours; they might be differentiated by use of dashed by dotted lines (Fig. 18.5d). Such contoured fault surfaces together with the structural surfaces they disrupt help define the geometry of the fault system, which in turn may lead to information concerning the type and amount of displacement.

Structure Contours and Complex Folds. Structure contours are a powerful method of expressing the form of gently warped structural surfaces. More complex forms may also be represented by such contours. There are two special cases. If the fold has a vertical limb, the contours will merge into a line that represents the map trace of that limb. If the fold is overturned, the hidden underside may be indicated with dashed contours.

Structure contours have also been used to study complexly refolded rocks (Berthelsen, 1960). As subsurface data in such areas are rare, the necessary information must be collected from surface mapping. Accurate measurements of the location and attitude of the surface of interest must be made. Clearly, being confined to the earth's surface, only a relatively small part of the chosen horizon may be confidently contoured. Great topographic relief extends this range.

FORM LINE CONTOURS

In some areas a recognizable key horizon may be lacking. If a structure contour map is to be drawn in such cases, it must be constructed from measured attitudes alone. Because no particular horizon is being represented, absolute values can not be given to the contour lines. However, the configuration of the structure can be shown by the pattern and spacing of the contour lines. Such lines without values relative to some datum are called *form line* contours.

For any given contour interval, the appropriate spacing between contour lines representing an inclined plane of known dip is given by

$$\text{spacing} = (\text{contour interval}) \cot \delta$$

In Fig. 18.6a the spacing for dip angles form

$1°$–$4°$, for a 5 m contour interval are shown. It is convenient to plot these spacings to the scale of the map along the edge of a card to speed the actual contouring. Given two or more dips along a line, the problem is to draw contours to quantitatively express the inclination of the surface (Fig. 18.6b).

PROCEDURE

1. Because the contours have no absolute value, they may start anywhere. In this simple example, contour a is arbitrarily located.

2. The next measurement in the down-dip direction is $1°$, suggestive that a $1°$ spacing be tried. It is found that this distance extends equally on both sides of the dip location, and therefore the second contour b is drawn. Note that this assumes that the average dip over this distance is $1°$. Since it ranges from approximately $0.5°$ to $1.5°$, and there is no evidence of an abrupt change, this must be nearly correct.

3. The next dip is $2°$. However, if a $2°$ spacing is tried, it is clear that the average dip over this distance is less then $2°$, though it is certainly more than $1.5°$. This obviously requires a spacing between $1°$ and $2°$, and closer to the latter. By trial and error a balance between the spacing and the average inclination is obtained; this can be done fairly accurately by judging the fractional spacing by eye.

4. This balancing process continues in this manner, until the required contours are drawn.

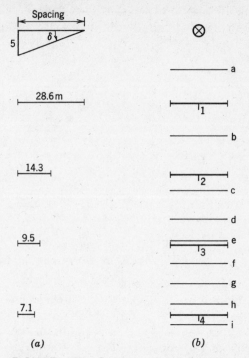

FIGURE 18.6 Spacing of form line contours.

For problems of contouring the attitude data on a map (Fig. 18.7a), proceeds in much the same manner. The initial contour is drawn with arbitrary location, but as closely parallel to the measured strikes as possible. It is often

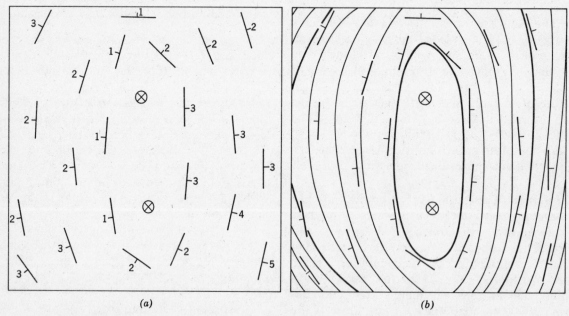

FIGURE 18.7 Form lines contours. (a) Map of measured attitude. (b) Final map. (After Badgley, 1959.)

found convenient to start by defining the highest part of the structure in this way. Subsequent contours are then drawn with a spacing which is appropriate to the local dips.

ISOPACH AND ISOCHORE MAPS

A closely related type of map is based not on the elevation of a certain horizons, but on the *thickness* of the unit. Such a map is called an *isopach map*, and the contours are lines of equal thickness. Isopach maps are very important in studying the regional thickness variations of sedimentary units as an aid to an understanding of the paleogeography at the time of deposition. A slightly different type of map is based on the vertical distance between the top and bottom of the strata in question. This measurement is the "depth" from the top to the bottom. Such a map is an *isochore map*. The importance of this type of map lies in the fact that it defines more accurately the present condition of the strata, rather than the thickness which is largely a function of its primary condition. Estimates of the volume of a given unit can be made more easily from an isochore map. Also, given an isochore map of a unit and structure contour map drawn on its upper surface, a second structure contour map can be constructed from the bottom of the unit by subtraction. If the dip angles are small, the isopach and isochore maps are, for practical purposes, equivalent. The procedure for drawing both types are based on an areal distribution of data points, and are identical with those described for structure contours.

EXERCISES

1. The subsurface data map of Fig. X18.1 (see p. X-9) gives the locations where the elevation of the top of a particular formation and the isochore interval (in brackets) were determined. Draw three contour maps.
    a. Structure contour map on the top of the formation.
    b. Isochore map of the formation.
    c. Structure contour map of the base of the formation

2. On Fig. X18.2 (see p. X-10) surface attitude measurements are shown; the area has negligible relief and no mappable horizon was found. Show the structure with form line contours.

19
Maps and Cross Sections

THE GEOLOGIC MAP

A variety of structural techniques have been described in previous chapters. In the main, the approach has been one of dissecting the geologic map and examining the parts. The map is, however, something more than the sum of these geometric parts, and it remains to consider some of the more collective features.

Properly done, the map is an exceedingly important tool in geology. The graphic picture it gives of the location, configuration and orientation of the rock units of an area could be presented in no other way. Essential as the map is, however, it is not without limitations, and if it is to be of maximum use these limitations must be fully understood. The most important point to realize is that

geologic maps generally record both observations *and* interpretation. In part, the element of interpretation is due to a lack of time and complete exposure; it is almost never possible to examine all parts of an area. If a complete map is to be produced, this lack then requires interpolation between observation points, and such interpolation is, to a greater or lesser degree, interpretive.

To distinguish between observation and interpretation several devices may be adopted. Most commonly, special symbols are used to identify several degrees of certainty in the location of lithologic contacts (Fig. 19.1). The choice of these symbols depends both on the ability to locate the boundary in the field and on the scale of the map. The rule of thumb is that a solid line is used if the contact is known and located to within twice the

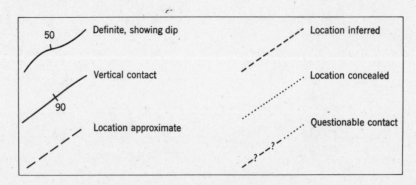

FIGURE 19.1 Map symbols for lithologic contacts.

164

width of the solid line in either direction (Kupfer, 1966). If contacts are drawn with the less certain symbols it is important that their inferred map positions make geometric sense; if a contact line crosses a valley it should obey the Rule of V's according to the inferred attitude. It is quite misleading to show uncertain contacts as if they were all vertical, though many examples of this practice can be found.

Factual and interpretative data may also be distinguished by considering the two aspects more or less separately. An outcrop map is one method of presenting field observations in a more objective fashion (Fig. 19.2a). Another way of conveying the essential information of an outcrop map, but without actually drawing in the boundaries of the exposed masses of rock, is to show abundant attitude symbols, which then serve two functions: to record the measured attitude, and to mark the localities where the attitude can be measured.

However, even an outcrop map or its equivalent can never be entirely objective, for several reasons. What constitutes an exposure of rock is itself subject to interpretation. For purposes of mapping, a thin rocky soil at the top of a low hill may be considered an outcrop by a worker in a poorly exposed area. In contrast, the geologist working in a mountainous region would probably not give such an exposure a second glance. Such differences

will certainly affect, and may even control the accuracy and completeness of the mapping. Furthermore, the identification of even well-exposed rock is not always so straight forward that all geologists would agree. Even with these limitations, it is, of course, still important to strive for as high a level of objectivity as possible, and to discuss the problems and difficulties involved in this quest in the text which accompanies the map.

There is another and much more fundamental reason why geologic maps are inevitably interpretive. Even the simplest rock mass is extremely complex, and a complete physical and chemical description of a single exposure might take years, and questions concerning the origin of the rock would almost certainly remain. Clearly such detailed studies are rarely feasible. The question then arises—what observations are to be made? The process of deciding what is important is guided in at least two ways. First, observations are made which have proved in the past to give results. Routine descriptions of attitude, lithology, visible structures and so forth are an important preliminary stage; some check lists have been published to facilitate this type of description. However, the creative part of field study involves asking critical questions and then attempting to find the answers. These questions are formulated on the basis of knowledge, intuition and imagination. In this search for understanding

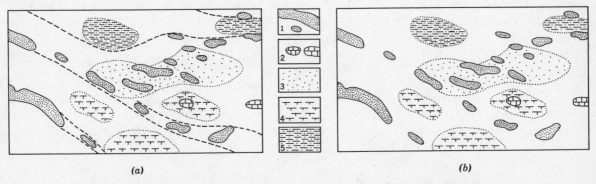

(a)   (b)

FIGURE 19.2 Hypothetical maps; lithologies are, 1 = sandstone, 2 = limestone, 3 = sandy soil, 4 = limy soil, 5 = clay soil. (a) Outcrop and soil map showing facies interpretation. (After Kupfer, 1966.) (b) Interpretation based on recognition of a mélange. The mélange blocks of sandstone and limestone are shown by outcrops and soils; the clay soil and covered areas are underlain by mélange matrix. (After Hsü, 1968.)

the older, often well established approaches may actually be a barrier which must be broken through if progress is to be made. For example, the interpretive aspect of the map of Fig. 19.2*a* is based on an application of the so-called laws of superposition, original horizontality, original continuity and faunal assemblage (see Gilluly, et al., 1968, p. 92, 103). There are, however, rock bodies which do not obey these laws; a *mélange* is an example (Hsü, 1968). French for *mixture*, the term mélange is applied to a mappable body of deformed rocks consisting of a pervasively sheared, fine-grained, commonly pelitic matrix, with inclusions of both native and exotic tectonic fragments, blocks, or slabs which may be as much as several kilometers long (Dennis, 1967, p. 107). Fig. 19.2*b* is an interpretive map based on the recognition of a mélange.

As with many things, progress in mapping is an evolutionary process. Each step along the way is an approximation. These approximations, because they are incomplete, necessarily involve some interpretation on the part of the investigator. Just as the making of the geologic map evolves, so too does the use of the map as an aid to understanding the structure and history of an area involve several stages. The first step does not constitute making structural interpretations, but is rather a repetition of the experience and thinking of the original observer. This step is indispensible in gaining a complete understanding of both the map and the area it represents, and the facility to do this can only be achieved by practice. Two attitudes toward

maps will greatly increase their usefulness:

1. Regard any map only as a progress report. Improvements can always be made by further work based on the original mapping, either by the study of new exposures, or a more detailed study with new concepts and techniques.
2. Develop a critical outlook toward the lines and symbols on the map. By refusing to accept them completely, especially those that are clearly interpretive, and by questioning the nature of the various structures and map units, new questions may arise that can be answered directly from the map, or from a visit to the area.

OTHER TYPES OF MAPS

Lithologic map units, and even different structural elements are often shown in color on geologic maps, in combination with the normal symbols printed in black. However, a carefully prepared black and white structural map is often superior to a colored one. On such maps the lithologic units should not be represented by purely geometric patterns, such as parallel rulings and other periodically reappearing patterns. Such patterns fail to express the variously curved lines of strike of the deformed rocks. It is both easier and conveys considerably more information to draw the strike lines freehand. Further, certain features can be depicted in this way which would be most difficult otherwise. For example, the transition between directionless and foliated rocks can be expressed by a parallel change in the map pattern. The two contrasting maps of Fig. 19.3 illustrates the value of this approach.

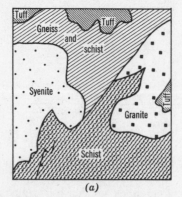
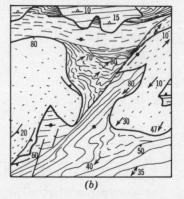

FIGURE 19.3 Two black and white structure maps of the same area. (From Balk, 1937, Geological Society of America. Used by permission.)

In addition to these surface maps there are a number of other types which may be constructed. Maps may, of course, be drawn wherever rocks are exposed, as in a mine. A structure contour map is a type of geologic map. Similarly, an isopach map is a geologic map, with the zero contour being the underground equivalent of a contact. An isopach map is also a picture of the structure of the lower boundary of a formation at the time the upper boundary was horizontal. A *paleogeologic* map portrays the formation immediately below the surface of an unconformity. A worm's eye map is a picture of the formations overlying the surface of the unconformity as seen from below; Levorsen (1960) gives a number of examples. *Palinspastic* maps restore the rock units to their relative original positions before structural displacement. Although difficult to draw, such maps are important becuase they introduce stages of historical development into the description of an area.

GEOLOGIC HISTORY

After describing the geometry of a rock mass, the next step is to work out the time sequence by which that geometry developed. This concern for history includes both the local chronology, and the dating of these events in terms of the geologic time scale. The dating is largely a matter of paleontology and radioactive studies. The local sequence of events, however, can be worked out without reference to absolute time.

There is a feature, not previously discussed, which is of great assistance in dating structures and period of structural movement. An *unconformity* is a surface of erosion or nondeposition that separates younger from older rocks. There are several important types (Dennis, 1967, p. 159). An *angular* unconformity is characterized by an angular discordance between the two sets of strata (Fig. 19.4a). In contrast, the *parallel* unconformity is marked by an evident break between two sets of parallel strata (Fig. 19.4b). A *nondepositional* unconformity is a surface of nondeposition; physical evidence of this surface may not be evident, and paleontologic evidence may be needed to demonstrate the time gap (Fig. 19.4c). A *heterolithic* unconformity describes the situation where the older rocks are non-stratified (Fig. 19.4d).

In determining the local chronology it must be kept in mind that several events may have been synchronous; for example, deposition may occur during folding and faulting. A further complication is that a given structure may be the result of several episodes of movement. Nevertheless, though it may be quite involved, the sequence of events can be worked out using rather simple geometrical relationships. The following criteria are self-evident:

1. *Folds* are younger than folded rocks.
2. *Faults* are younger than the rocks they cut.
3. *Metamorphism* is younger than the rocks it effects.
4. The erosion represented by unconformity is younger than the underlying rocks and older than the overlying ones. This is strictly true only for a small area; erosion and deposition at widely separated localities may be synchronous.
5. Intrusive igneous rocks are younger than the host rocks. This is especially clear where they are in cross-cutting relationships. A similar rule holds for other types of intrusions such as salt domes and

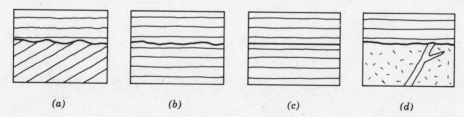

(a)   (b)   (c)   (d)

FIGURE 19.4 Important types of unconformities: (a) angular unconformity, (b) parallel unconformity, (c) nondepositional unconformity, (d) heterolithic unconformity.

sandstone dikes, with the qualification that the act of intrusion is younger though the material may be either older or younger.

An elementary, hypothetical map and the minimum number of events in chronological sequence derived from it illustrates the approach (Fig. 19.5).

THE STRUCTURE SECTION

One of the problems in reading a geologic map is to perceive the structures portrayed in two dimensions in their proper three-dimensional setting. Several techniques for doing this, especially the powerful down-structure method of viewing maps, have been described in earlier chapters. Vertical structure sections, though they have their limitations, are also useful in helping to work out and depict the structural relationships at depth, particularly when the structures are diverse and no single down-structure direction exists. The technique for constructing such vertical sections is straightforward and in general consists of two parts:

1. The topographic profile along the line of the section.

2. Structural data, such as contacts, attitudes, and so forth, which are added and then extrapolated into the underground.

The line of section is chosen to show the required structural features to best advantage. It is usually drawn as close to perpendicular to the strike direction as possible, although this is not always possible, or in some cases, even desirable.

Topographic Profile. The edge of a piece of paper is laid the full length of the chosen section line (Fig. 19.6). Points of intersection of the topographic contours and the section line are marked along this edge. Other features, such as the crests of hills or the locations of streams should also be marked, even though a contour line is not present. The values of the contours must be indicated.

A series of elevation lines are drawn on a second sheet of paper with a spacing equal to the contour interval and plotted to the same scale as the map scale. The topographic points along the section line are then transferred from the edge of the marked paper, which now represents the line of section, by projecting the contour marks to the corresponding elevation lines, and the series of points so formed is joined with a line representing the

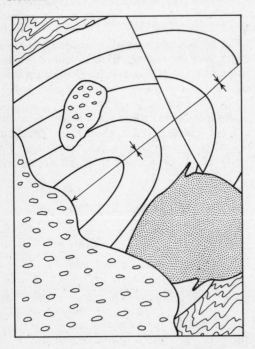

HISTORY

1. Deposition of pre-metamorphic sedimentary rocks
2. Folding and metamorphism of these rocks
3. Uplift and erosion
4. Deposition of a second sedimentary sequence
5. Folding
6. Faulting
7. Igneous intrusion
8. Erosion
9. Deposition of a third sedimentary sequence
10. Erosion to present topography

1   2   3   4

FIGURE 19.5 History from a geologic map.

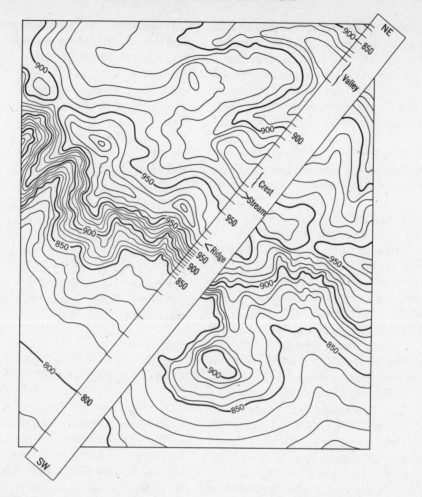

FIGURE 19.6 Line of section and construction of a topographic profile from a map.

topography. If the spacing of the contours is wide, the map may be consulted to assist in sketching topographic details (Fig. 19.7).

In constructing the topographic profile, it is an easy matter to exaggerate the vertical dimension by enlarging the vertical scale while keeping the map scale constant. The effect of such an alteration of the vertical scale produces profound geometrical changes, which

are examined in some detail in the next section. Generally, only unexaggerated sections should be constructed for serious structural work, because only with them are the true geometric relationships preserved. If more space is required to plot a wealth of data, the whole section should be enlarged uniformly.

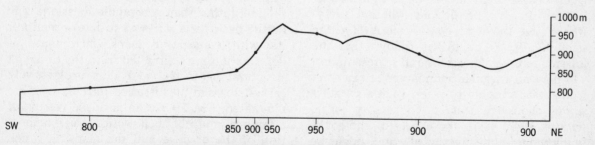

FIGURE 19.7 Topographic profile along the line of section.

Structure section. The second step is to add structural data to the constructed topographic profile. The various contacts and attitude points can be marked at the same time as the topographic data, although it is less confusing to make two separate plots, especially if there are abundant details which must be transferred.

Map data on either side of the section line may be projected to the section; the allowable distance of projections depends on the constancy of the trend of structural features normal to the structure line. Where the structures plunge, the map patterns may be projected to the plane of the vertical section, essentially the same construction used in constructing fold profile. When this is complete, all the surface information will have been utilized.

In extrapolating this surface data downward into the underground several different approaches may be used. In folded sedimentary rocks the projection may be done with the method of fold reconstruction by tangent arcs (Fig. 8.11). Lithologic boundaries with no regular features can only be projected downward using the surface attitude; this is only a first approximation and will, at best, be valid only for relative shallow depths. One example of such an unpredictable boundary is the contact of a discordant intrusive igneous body. Alternatively, such contacts might be shown diagramitically.

After taking these rather purely geometrical steps, the making of further predictions will depend on a thorough understanding of the various processes of folding, thrusting, faulting, and so forth, and this can only come with experience. As on the geologic map, the various lines of the structure section should indicate the degree of certainty in location. Questionable areas should be so indicated or left blank. The predictions will not everywhere have the same confidence level at the same depth. From this it follows that the lower limit of the structural representation will have an irregular boundary. For example, it might be possible to reconstruct accurately several thousands feet of sedimentary rocks; on the other hand, the nature of the rocks underlying a thin sheet of alluvium might be completely unknown.

If structure sections are prepared in black and white, lithologic symbols should be used to indicate the different rock units in much the same manner as with black and white structural maps. A variety of symbols are used, a few of which are shown in Fig. 19.8. The meaning of any such symbols used must, of course, be identified on the section, either by labeling the units directly or by including a list in the legend.

In its final form the section should be labeled with geographical coordinates because knowing the orientation is as important as its location. The section line should appear on the accompanying map. Prominant topographic features should be labeled to assist in orienting the reader. The scale, especially if different from that of the map, is also important.

OTHER TYPES OF SECTIONS

A number of variations are possible. Composite sections can be drawn, by projecting data to the section plane from some distance or by combining in one section several different lines that meet at angles. This is generally done to show diversity, which would not be possible along a single line of section.

One way of suggesting three dimensions is to use multiple sections. Two groupings are common, but any number of combinations are possible. A *coulisse* diagram is a group of parallel sections drawn and arranged serially to take advantage of some special point of view, such as along the strike of a fault, or in the direction of a fold axis. *Fence* diagrams may be thought of as two intersecting coulisse diagrams giving appearance of an egg crate.

Time, rather than geographic location, may be the basis for a series of structure sections. As with palinspastic maps, increments of deformation are subtracted from the observed structural geometry, and thus progressively earlier stages in the historical development are illustrated. Fig. 19.8 is an abridged version of a famous group of such sections; an examination of the original, and the maps on which they are based is well worth while.

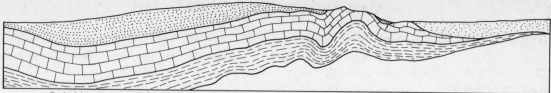

Early folding near margin of sedimentary basin with simultaneous deposition of coarse clastics in the marginal trough.

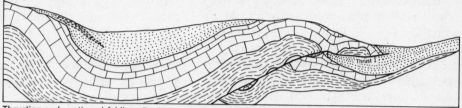

Thrusting and continued folding. Rocks are carried toward the trough. Deposition of coarse clastics continues.

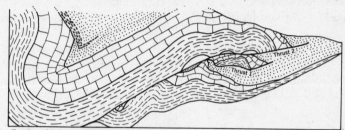

Further folding with formation of new thrust.

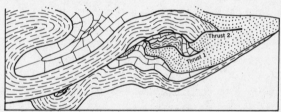

Thrusts are folded, as main syncline becomes recumbent.

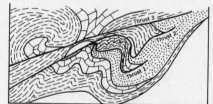

Involution of the syncline and renewed thrusting.

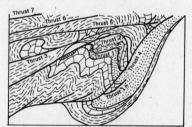

Tightening up all of folds and final imbricate thrusting.

Coarse clastics sediments: sandstones and locally conglomerates

Limestone

Shale

Basement rocks (greenstone, breccia and cherts)

FIGURE 19.8 Diagramatic sections showing progressive development of complex folds and thrusts. (After Ferguson and Muller, 1949.)

VERTICAL EXAGGERATION

It is occassionally useful, and still a very common practice, to draw cross sections with the vertical scale enlarged relative to the horizontal scale; that is, to stretch the section vertically while leaving the horizontal dimension unaltered (see Suter, 1947). This practice is almost invariably followed in stratigraphic or geomorphic sections where more space is needed to plot vertical details or to accentuate certain features which would otherwise be obscure. The result is known as a vertically exaggerated cross section, and the degree of the stretch is defined by an exaggeration factor V, where

$$V = \frac{\text{vertical scale}}{\text{horizontal scale}} \qquad (19.1)$$

For example, if the horizontal scale is 1/50,000 and the vertical scale is 1/10,000, the exaggeration factor is 5.0.

Largely because of continued exposure to such sections most geologists tend to think in terms of them, and unexaggerated sections often have an "unnatural" appearance. It is therefore vital to understand the detailed geometric implications of vertical exaggeration. This will aid both in deciding whether to draw such sections or not, and in interpreting the exaggerated sections of others.

As a result of this stretching, both the angle of dip (or slope) and thickness are systematically distorted in the exaggerated sections (see Fig. 19.9). A fruitful approach to describing these changes is to consider vertical exaggeration as artificially introduced strain. Not only does this allow previous results

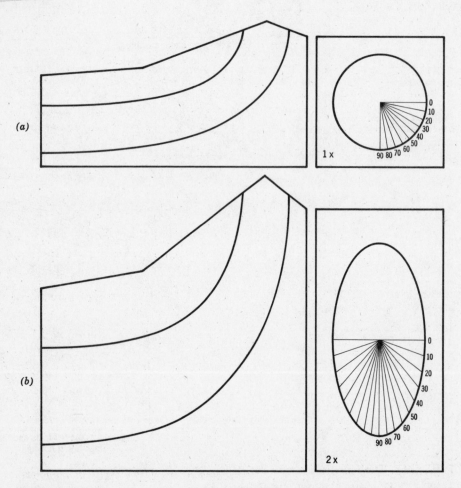

FIGURE 19.9 Natural and vertically exaggerated structure sections. (In part after Wentworth, 1930.)

based on strain theory to be used, but it also emphasizes the profound effect which accompanies the exaggeration.

The relationship describing the change in orientation of any line during strain is given by (6.6). Noting that for a line of dip, angle $\phi = (90 - \delta)$ and $V = R_s$, and using the identity $\tan(90 - \delta) = 1/\tan \delta$, then

$$\tan \delta' = V \tan \delta \qquad (19.2)$$

where δ and δ' are, respectively, the original and the exaggerated dip. In Fig. 19.10 this equation has been solved for selected values of δ and for values of V ranging from 1 to 20. An examination of this graph makes clear the effect of vertical exaggeration on the dip angle. In general all dips are steepened, but small dip angles are affected relatively more,

with the result that differences in inclination are accentuated. It is this property which is of advantage in depicting subtle variations on the longitudinal profile of a river. On the other hand, differences between steep dips are minimized. For example, with a vertical exaggeration of $V = 10$, lines dipping at angles of 30° and 60° will appear with inclinations of 80° and 87°. Thus the mechanically important distinction between a normal fault and a thrust fault would be virtually lost, and this is a serious disadvantage.

The exaggerated thickness t' is given directly by (6.9). Fig. 19.11 gives the solution to this equation for values of V from 1 to 20, assuming an initial unit thickness. As can be readily seen, the exaggerated thickness varies according to the original dip. The limiting

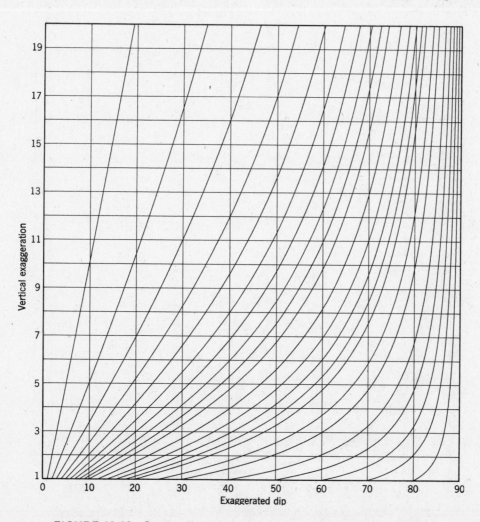

FIGURE 19.10 Graph of angle of dip (or slope) and vertical exaggeration.

cases occur when the layer is horizontal and the thickness is multiplied by the exaggeration factor ($t'_0 = Vt$), and when the bed is vertical and the thickness remains unchanged ($t'_{90} = t$). Most of the variation in exaggerated thickness occurs in a rather narrow range of original dips. For example, if $V = 10$, then $t'_0 = 10$ and $t'_{25} = 2.3$. Thus a layer of uniform thickness but variable inclination within these limits will appear to have a four-fold variation in thickness; it is easy to see that small real variations in thickness would be masked. As an aid to interpreting exaggerated sections, a more useful form of this graph is to plot exaggerated thickness against exaggerated, rather than original dip (Fig. 19.12).

From these considerations it should be clear that vertically exaggerated sections distort the form and orientation of the structures, thus tending to destroy the very information the structure section seeks to show—the true geometric relationships at depth. Therefore they should not be used for serious structural work. For those few situations where a vertical exaggeration may aid a presentation, the smallest possible exaggeration factor should be used. In addition, there is an important responsibility to make the reader well aware of the degree of the exaggeration. This can be accomplished in several ways: (1) include bar scales for both the horizontal and vertical dimensions, (2) quote the actual exaggeration factor, (3) include a protractor of exaggerated dips, and (4) include a natural scale section in addition to the exaggerated one for easy comparison. This latter is perhaps the most effective method, but it is quite useful to

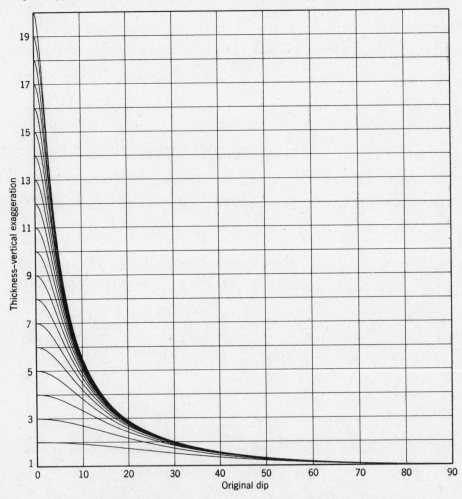

FIGURE 19.11 Graph of thickness and original dip angle.

supply all the items. Since exaggerated sections are so common, a natural scale section should be clearly labeled *no vertical exaggeration*.

There is another subtle distortion which appears when sections of regional extent are drawn as if the earth were flat. Sea level is depicted as a horizontal straight line, so that the floors of sedimentary basins appear distinctly concave upward; vertical exaggeration further compounds the distortions. If true sections are drawn, it is, of course, sea level which appears as a curve and the basin floor more nearly approximates a straight line. This difference in basin geometry has important bearing on the mechanical properties of the basin fill (Price, 1970, p. 15), and on the mechanics of basin evolution (Dallmus, 1958). Again, construct accurate, true sections. Then if there is reason to distort them do so with caution, and advise the reader of what you have done.

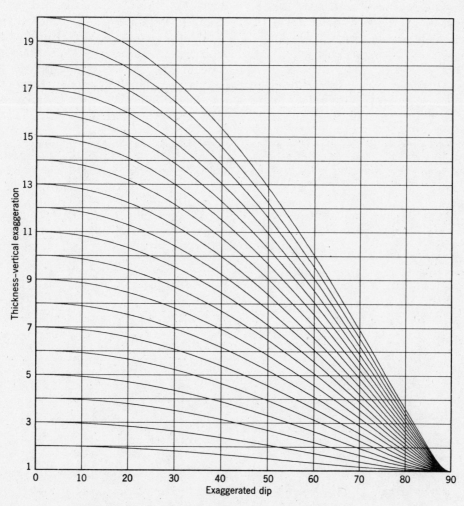

FIGURE 19.12 Graph of thickness and exaggerated dip angle.

EXERCISES

1. On a published section depicting folds in the Michigan Basin, a unit distance on the vertical scale represents 200 feet (= 61 m), and on the horizontal scale this same distance represents 7 miles (= 11.25 km). The steepest dip shown is 73°. What is the vertical exaggeration factor, and what is the real dip angle? What do you think about this degree of exaggeration, and what would you suggest as an alternative?

2. Using an available map, construct a true scale cross section showing both topography and structure.

20
Block Diagrams

The block diagram is one of the best ways of presenting a wealth of geologic information in a compact, three-dimensional form. Almost at a glance, the relationships between the data plotted on the visible surface of the block can be integrated into a spatial picture. The construction of such a diagram entails drawing a scaled block, adding topography to the upper surface, and representing the geologic structures on its sides. Scaled blocks may be constructed by descriptive geometry, but the method is fairly involved and time consuming, especially in the light of the alternatives. With the aid of the orthographic net (Fig. 20.1), a scaled cube may be constructed directly in any predetermined orientation. Blocks of any other shape can then be assembled by grouping a number of such cubes. The choice of orientation of the cube depends on the desired relative prominance of the three visible surfaces, or on the orientation of some particular structure, such as the axis of a fold, or the plane of a fault.

UNIT CUBE

The construction of the unit cube may be accomplished in two equivalent ways: (1) by revolving the cube into the desired position, and (2) by a direct plot. Because it aids visualization, the first method will be used to introduce the use of the orthographic net.

CONSTRUCTION BY ROTATION (Fig. 20.2)
1. A cube drawn with one side in normal view appears as a square (Fig. $20.2a_1$). The labeled corners are: B is the rear, upper righthand corner; A, C and O are the front three corners. With the orthographic net oriented with the 0-180° diameter oriented east-west and an overlay in place, OB is represented by a point at the center of the net, and OA and OC by horizontal and vertical radii, respectively (Fig. $20.2a_2$). The exact circumference of the net is easily found by counting 90° from the center.
2. Rotating through an angle a about edge OC exposes to view the originally hidden right-hand side of the cube (Fig. $20.2b_1$). On the orthographic net, this rotation moves points A, and B to the right, a distance equal to a (here $a = 40°$).

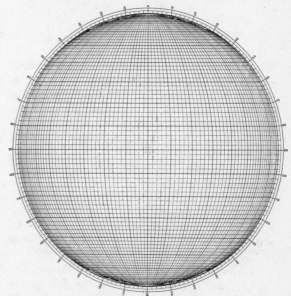

FIGURE 20.1 Orthographic net. (From Wright, 1911. Used by permission of Carnegie Inst. of Washington.)

177

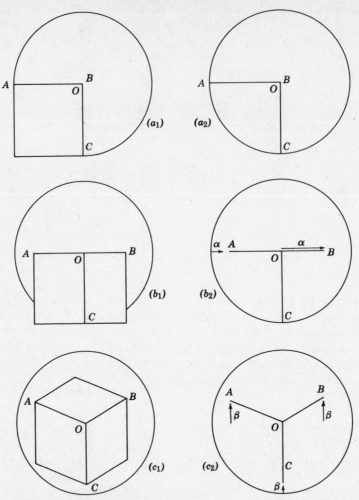

FIGURE 20.2 Construction of a unit cube by rotation with the aid of the orthographic net. (After McIntyre and Weiss, 1956.)

3. A second rotation through an angle of β about an east-west axis passing through point O exposes the top of the cube (Fig. 20.2b_1). On the net the points A, B and C move upward along the straight grid lines a distance equal to β (here $\beta = 30°$).

4. The three lines in this final position on the net represent the solid angle made by the front three edges of the cube, and they appear in their correct foreshortened proportions.

DIRECT PLOT (Fig. 20.3)

1. With the net oriented with the 0-180° axis east-west, the four points representing the three edges are plotted directly:

    a. Point O is fixed at the center of the net.

    b. Point C is located a distance β from the bottom of the net along the north-south diameter.

    c. Point A lies on the β great circle (measured from the center) and distance a from the left side.

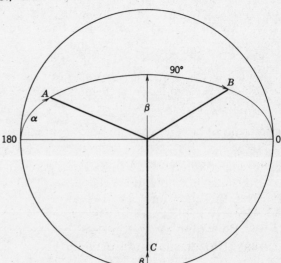

FIGURE 20.3 Unit cube by direct plot on the orthographic net. (After McIntyre and Weiss, 1956.)

d.   Point *B* lies 90° from *A* long this same great circle.

2. A comparison of the final result with that obtained by rotation shows that they are identical.

With a correctly foreshortened cube in the required orientation, a rectangular block may be built up by multiplying the unit dimensions along the three principal directions; Fig. 20.4 shows such a block. Fractions of the cube's dimensions may also be obtained by proportionally dividing one or more of the three directions.

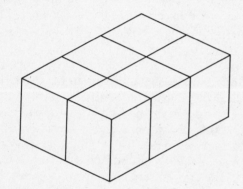

FIGURE 20.4   Block diagram by assembling units cubes.

TOPOGRAPHY

The upper surface of the block can be made to represent any given subsurface level, or even a generalized, planar part of the earth's surface. If the area has even a small amount of relief, the three-dimensional aspect of the block is greatly enhanced by adding topographic relief to the diagram. A number of systems have been devised, including pantograph-like drafting machines, to adjust map topography systematically to the proportions and scales of the block diagram. This topography can also be drawn directly by using a relatively simple graphic method. Given a topographic map, or any part of it, the problem is to show the surface land forms on a block of any desired orientation.

CONSTRUCTION (Fig. 20.5)

1. Draw a square grid on the map with the ordinate in the direction of the proposed line of sight. The actual grid spacing should be dictated by

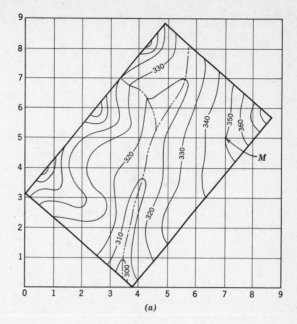

(a)

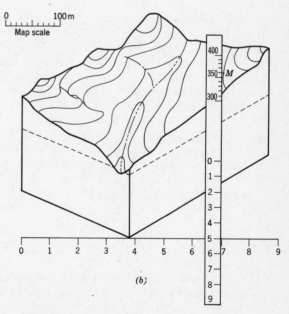

(b)

FIGURE 20.5 Topography on a block diagram. (*a*) Topographic map with superimposed grid. (*b*) Method of transferring topographic detail to the block. (After Goguel, 1962.)

the amount of detail to be transferred to the block. (Fig. 20.5*a*).

2. Draw a unit cube in the required orientation. Position this cube below the map so that its front corner lies exactly along the line of sight to the front corner of the map. The cube can then be multiplied to the dimensions of the map by drawing

other line-of-sight lines from the outside corners of the map (Fig. 20.5*b*).

3. The depth of the block depends on the depth of structure to be shown. In the example, the 300 m level is placed at the top of the unit cube.

4. Along the base of the block reproduce the abscissa of the grid and in the correct position with respect to the map grid.

5. From an oblique view, the front-to-back grid spacing is foreshortened; this foreshortened grid scale is related to the map grid scale by a factor of sin β, where β is the vertical angle the line of sight makes with map plane (cf Fig. 10.10). This corrected scale is plotted along the edge of a strip of paper.

6. The map scale, as measured vertically, is similarly reduced by a factor of cos β. This new scale is also added to the strip of paper. The spacing between the two scales depends on the distance between the abscissa scale and the top of the unit cube, and a little experimentation will show the correct placement.

7. The positions on the block of a series of topographic points (or any other points) are then located. For example, point M is at the corner of the horizontal grid number 5 and vertical grid number 7. On the block, 5 on the corrected grid scale is moved to the 7 on the lower abscissa scale, keeping the measuring strip vertical. The elevation of M is 350 m and this point is found on the corrected vertical scale and the point plotted. The procedure continues until enough points have been located. The topography may then be shown as foreshortened contours (Fig. 20.5*b*), or by using a simple shading technique (Fig. 20.6).

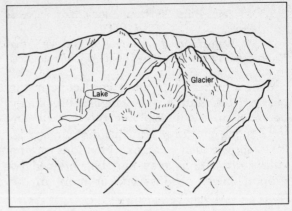

FIGURE 20.6 Topography rendered by simple shading. (After Lobeck, 1958.)

transfer geologic data from the map in the same manner.

For the vertical sides of the block, a different approach must be used. First, two vertical sections are constructed along the lines representing the two edges of the block. Fig. 20.7 shows one such section with a superimposed grid. Next, this same grid is drawn on the corresponding side of the block using the appropriate foreshortened scale. The structural data is then transferred to the block.

Various lithologic units can be shown with appropriate symbols on both the map and sections. On the map surface they should be kept sparse so the topographic details will not be obscured.

STRUCTURAL DATA

Once the topography surface of the block has been constructed, it is a simple matter to

MODIFIED BLOCKS

In order to show certain features better, a number of modifications may be used. The

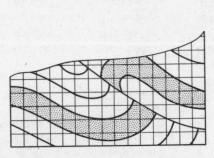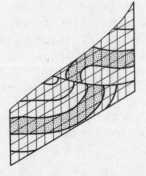

FIGURE 20.7 Plot of structural details on the side of a block by transferring to a distorted grid.

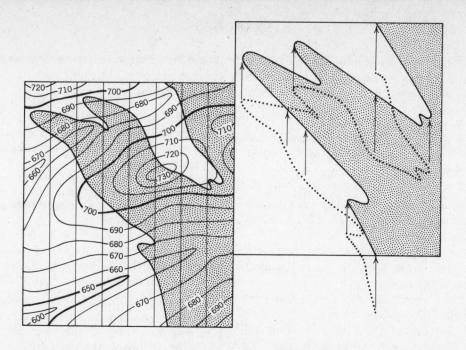

FIGURE 20.8 Method of projecting map data to a horizontal plane surface. (From Turner and Weiss, 1963, Structural Analysis of Metamorphic Tectonites. McGraw-Hill. Used by permission.)

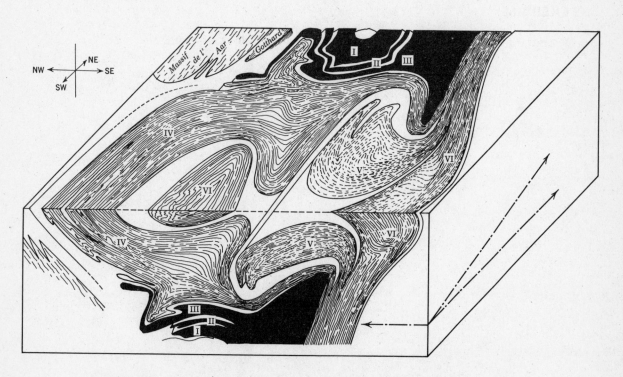

FIGURE 20.9 Block diagram showing plunging structures in the Alps. (From Argand, 1911, Beiträge zur geologischen Karte der Schweiz. Used by permission.)

block may be cut into pieces and the pieces separated to expose its internal parts. Similar cuts may be made to remove corners or variously shaped slices to show other structural details to advantage.

Another way of emphasizing certain features is to dissect the block along structural surfaces. For example, a complexly folded and faulted stratigraphic horizon could be shown by artificially removing all of the overlying material; Goguel (1962, p. 134) shows such a diagram. A variation would be to dissect several such horizons in the same structure and show them in an expanded vertical stack.

Especially in mountainous areas, the presence of relief on the block may hinder rather than aid the presentation, and it may be desireable to eliminate the complications of outcrop pattern caused by it. This can be accomplished by projecting the structures to a horizontal plane. Any plane can be used, but

it is often convenient to use sea level because the topographic contours also use this elevation as datum.

PROCEDURE (Fig. 20.8)

1. On a transparent overlay rule a series of closely spaced lines parallel to the trend of the fold axes on the geologic map.

2. Select a series of points on the contact of a lithologic marker in the fold. These points should be spaced closely enough to allow the structure to be accurately sketched in.

3. Each point is projected to sea level by moving it parallel to the trend lines in the direction of the plunge, through a distance $= h/\tan p$, where h is the elevation of the point and p is the angle of plunge.

Fig. 20.9 is a famous block diagram with an artificially planar map surface that shows the plunging structures of the Pennine Nappes of the Alps. Although the axial continuity of the individual nappes is not as persistent as formerly thought (Rutten, 1969, p. 195), the method remains valid.

EXERCISE

From an available geologic map, construct a scaled block diagram.

APPENDIX
A
Elements of Descriptive Geometry

Descriptive geometry is the art of accurately drawing objects, and of graphically solving space problems.

GRAPHIC REPRESENTATIONS

Projections. The portrayal of a three-dimensional object in two dimensions is accomplished by projecting the various parts of the objects form to a plane, and the resulting representation is a *projection*. Photographs, maps, various drawings, and shadows are common examples of projections. Photographs and shadows are the result of projecting the various parts of the object by light rays. These rays are *projectors* that join the parts of the object with the corresponding parts of the projection on the *image plane*. Similar projectors and projections can be mechanically constructed by a fairly large number of different methods.

Orthographic projections. One of the simplest methods of projection, and the one used almost exclusively in engineering drawing, as well as for many purposes in geology, is called the *orthographic* projection. Orthographic means "drawn at right angles", and refers to parallel projectors that are perpendicular to the image plane. An everyday example is the shadow of an object on a wall that directly faces the sun caused by the

parallel rays of sunlight; such shadows are exactly the same size and shape as the object.

These definitions relating to orthographic projections are illustrated in Fig. A.1. As in this example, it is often convenient to refer to three separate image planes: one horizontal (called the plan or map view) and two vertical planes at right angles to each other. These three are the *principal* views. However, the number and position of these planes is not fixed, and other image planes, giving *auxiliary*

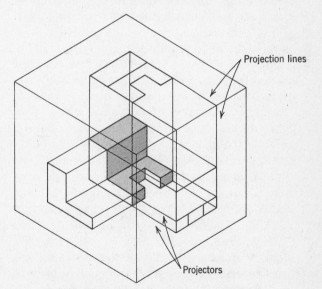

FIGURE A.1 Terms used in orthographic projection. (After Warner and McNeary, 1959.)

183

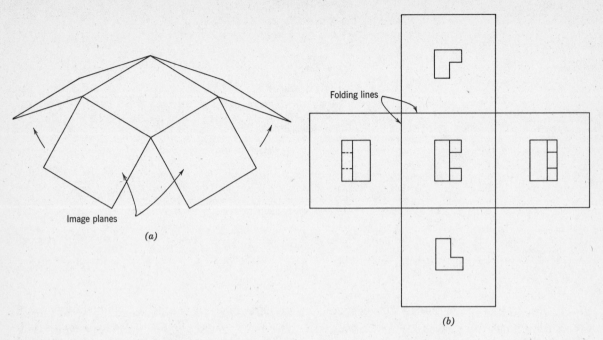

(a)

(b)

FIGURE A.2   Unfolding of the cubic box of Fig. A.1.

views, inclined at any angle may be used and in fact may be required to describe the object thoroughly. In Fig. A.1 the object is projected orthographically to the sides of the cube, and the whole then projected to the plane of the paper. In order to show the true size and shape of the several views of the object it is necessary to obtain a direct view of the image plane, that is, obtain a line of sight which is perpendicular to the image plane. This may be accomplished by unfolding the cube just as one unfolds a cardboard box, so that all the image planes lie in a single plane. Fig. A.2 shows the unfolding of such a cubic box. The edges of the cube which act as hinges during this flattening are termed *folding lines* (abbreviated *FL*). Any edge may arbitrarily be picked to act as a folding line, although it is often simpler to use folding lines in the horizontal plane. If auxiliary image planes are used it is necessary to rotate about other folding lines, possibly through angles other than 90° during this unfolding operation.

In practice, of course, the three-dimensional box is never actually constructed nor is the unfolding process literally followed. These steps are by-passed and the needed orthographic views are constructed directly. Fig. A.3 illustrates the construction with the use

of lines connecting corresponding points on two or more image planes drawn across the folding lines; these lines are called *projection lines.*

Visualization in three-dimensions.   The purpose of orthographic projections is to describe in the simplest manner possible the actual dimensions, shape, and position of an object in three-dimensional space. The process may also be reversed. Given a series of image planes it is important to develop the ability to visualize the object in its three-dimensional setting. To do this, two or more views must be recombined.

Visualization of points.   The projection of a point in space is always another point. Two views are sufficient to establish the location of the point; this location can then be described with respect to another point, or to a line or plane, or any combination (Fig. A.4).

Visualization of lines.   The projection of a line in space is generally another line. The length of the projected line is equal to the length of the original line only if the image plane and line are parallel. When they are not parallel, the length in projection decreases as

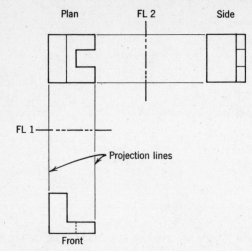

FIGURE A.3 Construction of the three principal views using folding lines and projection lines.

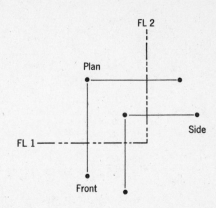

FIGURE A.4 A point in space in orthographic projection.

the angle increases until an image plane is perpendicular to the line and the line appears as a point (Fig. A.5).

Visualization of plane figures. The projection of a plane figure to an image plane parallel to itself represents the true shape of the figure; this is called the *normal* view because the line of sight is then perpendicular to the plane of the figure. A plane figure projection to an image plane perpendicular to itself gives a straight line; this is an *edge* view. Intermediate views show neither true size nor shape, but do give more information on the form of the figure than does an edge view (Fig. A.6).

Visualization of solid figures. The visualization of plane-bounded solid involves visualizing the position, size and shape of the individual plane figures and the lines and points of intersections of these planes. Fig. A.7 gives a few simple examples. Solids involving curved bounding surface may be similarly visualized, although it is usually more difficult.

Pictorial projections. The use of several views in orthographic projections is the only completely accurate method of fully representing complex space relationships. Only in this way are true distances and angles depicted. There are, however, other methods of presenting three-dimensional objects. While

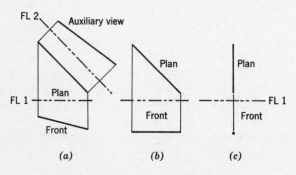

FIGURE A.5 Lines in projection: (a) inclined line, (b) horizontal line, (c) horizontal line in end view.

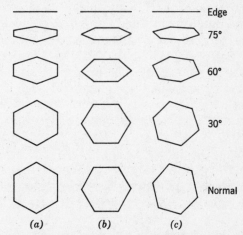

FIGURE A.6 Regular plane hexagons viewed in various rotations.

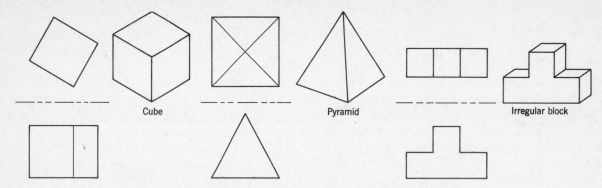

FIGURE A.7 Visualization of simple solid figures from two principal views.

not exact, they have the advantage of more direct tridimensional qualities. Because of the ease of visualization, they are termed *pictorial* projections.

Isometric projections. If a cube is placed with three visible sides equally oblique to the line of sight and the figure then orthographically projected to the plane of the paper, an *isometric* or "same-measure" projection results (Fig. A.8*a*). This position is attained when the edges of the three visible faces appear to the 120° apart. In such a projection no side is represented in its true length, but all sides are distorted equally. Thus, equal lengths on the cube are uniformly reduced by a factor of 0.8165. This property of equal lengths makes it a very useful type of projection.

Dimetric and trimetric projections. For reasons of emphasis or general pictorial effect, a different line of sight may be required. Depending on the exact position of this line of sight, several different scales may be

needed for the various directions, in contrast to the one for isometric diagrams.

If two of the axes are equally inclined and the third is unequally inclined to the projection plane, then two different scales are needed, one for the unequal and one for the two equal axes. The figure is said to be drawn in *dimetric* projection (Fig. A.8*b*).

If all three axes have different inclinations to the projection plane, then three different scales are needed, and the figure is shown in *trimetric* projection (Fig. A.8*c*).

Perspective projections. Block diagrams in isometric and related projections are readily understandable and because they maintain their proportions along the several axes, they are fairly easy to construct to scale. However, because objects appear of equal size regardless of their distance from the observer, objects appear to become bigger in this distance and this is most unrealistic. This defect may be corrected with the use of perspective projection.

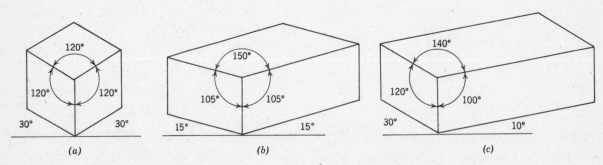

FIGURE A.8 Pictorial projections: (*a*) isometric projection, (*b*) dimetric projection, (*c*) trimetric projection.

Unlike the orthographic projectors that are perpendicular to the image plane, the law of perspective requires that the projectors pass through a single *station point*. Photographs are essentially perspective projections with the center of the camera lens being the station point. Because of the convergences of projectors, distant objects occupy smaller angles and thus appear to be smaller. Similarly, parallel lines converge toward vanishing points on the horizon.

Artists acquire the ability to draw correct perspective with practice. While they may use a straight edge to establish the horizon and to draw in the edges of rectangular objects, they most often judge correction proportions by eye. For most purposes this will give perfectly satisfactory results. Purely geometrical systems for constructing perspective projections have also been devised. One of the simplest is described by Doblin (1956), and his rationale is highly recommended for a more thorough understanding of the general principles of perspective. In his system the cube is the basic perspective form because it is the simplest rectangular figure which height, width and depth appear in their correct relative proportions, and because other forms may be built up using a number of cubes. Perspective block diagrams are much more work to draw, but outstanding results can be obtained which are well worth the effort.

CONSTRUCTION OF ELLIPSES

The need for ellipses arises in a variety of situations; one of these is closely associated with the pictorial projections. If a circle is inscribed in the unit square of one or more faces of a cube, and this cube is then drawn in isometric, dimetric or trimetric projection, the square will appear as a parallelogram, and the circle as an ellipse. There are templates which allow a series of sizes and shapes to be traced out directly. The various shapes are graded according to the angle the line of sight makes with the oblique plane of the circle; ellipses for every 5° interval over the range of 10° to 60° are commonly available. There are

also several devices which allow a complete range of ellipses to be drawn mechanically; "Rulalipse" and the "Ellipsco Compass" are two examples.

Any ellipse may also be constructed by a simple method. The two axes of the ellipse are drawn in the required orientation, and the appropriate axial lengths indicated. On the edge of a strip of paper, mark off the lengths of the semiaxes such that AC is the length of the semimajor axis and BC is the length of the semiminor axis. By moving the strip of paper while keeping the marks A and B on the appropriate axes, the point C traces out an ellipse (Fig. A.9).

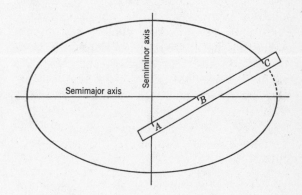

FIGURE A.9 Construction of an ellipse.

GRAPHIC SOLUTIONS

Graphic solutions of space problems are not absolutely accurate. Limiting factors include the scale of the drawing, accuracy of the instruments, and the skill of the draftsman. For example, with care it is possible to draw a pencil line as narrow as about 0.1 mm. The point of intersection of two mutually perpendicular lines is then 0.1 mm in diameter, and thus the probable error in measuring the distance between two such points is about 0.1 mm, which becomes progressively greater as the angle between the intersecting lines becomes more oblique. As the accuracy of the line is independent of scale, the accuracy of a given graphic solution depends linearly on the scale of the drawing, other things being equal. If the scale is doubled, the percentage error is one half.

Theoretically, a mathematical solution is capable of absolute accuracy. But no solution can be more accurate than the original empirical data, and therefore a graphic solution is just as accurate so long as it is within the limits of the numerical observations. In addition, many graphic solutions are quicker, and because they directly involve three-dimensional visualization, they are more readily understandable and may yield relationships otherwise hidden.

In the final analysis, the choice of method, whether mathematical or graphical, depends on the requirements of the problem and the observational data. In geology great accuracy of the data is difficult to attain. Thus, graphic solutions produced under normal working conditions, using average quality instruments and observing appropriate construction techniques, are sufficient for most purposes. The method to be used and the necessary accuracy, however, must be decided for each problem.

In any case, the accuracy of a graphic solution may be improved in the following ways:

1. Enlarge the scale. The optimum scale is just slightly larger than the data requires. This insures requisite accuracy and economy of time.
2. Draw the lines as narrow as possible with a hard, sharp pencil, and with light pressure.
3. Locate intersection of lines with the use of an angle as near 90° as possible.
4. Measure angles with a large radius protractor.
5. Avoid cumulative errors. If possible, measure off the total length of a line rather than step off the distance with a divider; or measure the total length without lifting the scale for intermediate points.
6. The use of high quality instruments maintains a higher degree of accuracy.
7. If the drawings are to be worked on over a considerable length of time, dimensionally stable materials should be used. Often a more practical approach is to finish the construction in as short a working time as possible.
8. Mistakes may be minimized by keeping the actual construction simple and compact, and by labeling all the points on the drawing.

Concerning the space relations of points, lines and planes, there are just seven basic *metrical* problems, that is problems involving the measurement of distances and angles.

They are:

1. Length of a line segment.
2. Angle between two lines.
3. Angle between two planes.
4. Angles between a line and a plane.
5. Distance from a point to a plane.
6. Distance from a point to a line.
7. Distance between two noncoplanar lines.

However, all seven can be solved with just two fundamental construction techniques: (1) the determination of the true length of a line, and (2) the rotation of a plane figure to any desired orientation. In the illustrations that follow, these two constructions are described in detail and then the various metrical problems are solved with one or both of these basic procedures.

The actual construction used in these examples is only one of several such techniques available. The details of the others can be found in any book on descriptive geometry. It should also be noted that two modifications have been used in the illustrations that produce inaccuracies, and therefore should normally be avoided: for emphasis certain lines are much too heavy, and, in general, the constructions have been expanded to avoid the confusion of too many overlapping lines.

The following sections should be reviewed in sequence. The two basic techniques are very important because all the other problems are solved with them. The separate problem solutions should not be memorized. Rather, strive to analyze the given problem by visualizing what is needed. Any actual construction is correct so long as it gives the right answer with appropriate accuracy.

True length of a line. To determine the true length of a line segment, a normal view of the plane containing the line must be obtained. There are two cases:

1. The line is oblique to two of the principal views, and parallel to the third. This third view shows the line in true length.
2. The line is oblique to all three principal views. The construction of an auxiliary view is required.

PROBLEM (Case 1)

Given the projection of an inclined line in plan

and side view construct the true length of the line and its inclination in the front view (Fig. A.10a).

APPROACH

Clearly, the projection of the line segment is fixed in the front view by the projectors connecting the line to this image plane. Because the projectors can not be constructed directly, we must seek an indirect method of obtaining equivalent information. As shown in Fig. A.10b the projection lines drawn from the plan view to the front view, and from the side view to the front view supply this information. This will be even more apparent if the rectangular box is unfolded (Fig. A.10c); note that since the front and side views are no longer in contact after unfolding, the projection lines are shown bridging this gap by circular arcs. It is a simple matter to construct the equivalent of this unfolded box directly.

CONSTRUCTION (Fig. A.10d)

1. Draw FL 1 perpendicular to the projection of the line in the plan and side views, and FL 2 parallel to the projection of the line.
2. Draw projectors from the plan to the front view perpendicular to FL 2. Similarly, draw them from the side to the front view perpendicular to FL 1; the circular arcs are constructed using the intersection of FL 1 and FL 2 as center.

3. The intersections of the two pairs of projection lines on the front view fixes the projection of the line on this plane.

ANSWER

Distance XY is the true length of the line, and θ is the angle of inclination measured from the horizontal.

Even if the actual object is more complicated than a single line segment, it can readily be seen that from any two principal orthographic views, the third may be constructed by this method.

PROBLEM (CASE 2)

Given the projection of an inclined line in plan and side view construct the true length of the line and its inclination (Fig. A.11a).

APPROACH

The method is virtually identical. Instead of the front view, an auxiliary view is constructed which serves the same purpose.

CONSTRUCTION (Fig. A.11b)

1. Draw FL 1 parallel to the plan-side view edge, and FL 2 parallel to the trace of the line in the plan view.

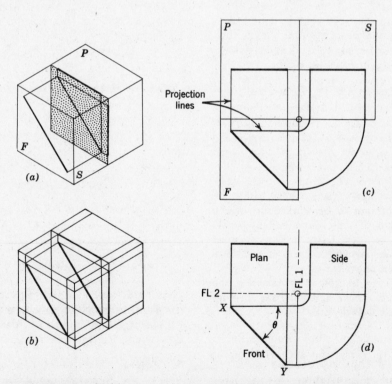

FIGURE A.10 True length of line parallel to one principal view.

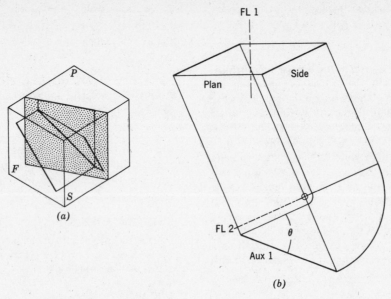

FIGURE A.11 Length of a line not parallel to any principal views.

2. Projection lines are drawn from both the plan and the side views perpendicular to FL 2. Again, the intersection of the two folding lines is the center of the circular arcs.

3. The intersections of these two pairs of projection lines establishes an auxiliary view (Aux 1) in which the true length of the line and its inclination appears.

Thus, given any two principal views, not only the third principal view, but any view may be constructed by this method.

Rotation of a plane figure. The rotational problem has two forms. The first, and simpler one, involves the rotation of a plane figure of known shape about a given axis into some desired orientation.

PROBLEM

Rotate the equilateral triangle *ABC*, shown in normal view, 60° about the given axis (Fig. A.12).

CONSTRUCTION

1. Draw FL 1 perpendicular to the given rotational axis to establish a front view.

2. Draw projection lines from the points *A*, *B*, and *C* perpendicular to FL 1. This establishes the edge view of the original triangle.

3. With the point of intersection of the axis and FL 1 as center, continue the three projection lines with 60° arcs. This gives the edge view of the rotated triangle.

4. Draw projection lines from this edge view back to the plan view. The intersection of these with

lines connecting corresponding points of the original triangle drawn perpendicular to the axis fix the corners of the rotated triangle, which can then be completed.

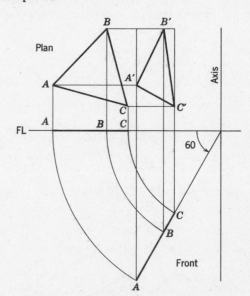

FIGURE A.12 Rotation of a plane figure.

The rotation of a plane figure into a normal view is more involved, because the orientation of the rotational axis is not initially known, and must be found.

PROBLEM

Rotate the triangle *ABC*, as depicted in the given front and plan views separated by FL 1 into a normal view, and thus determine its true shape (Fig. A.13).

APPROACH

In order to identify the rotational axis it is necessary to draw a view in which one side of the triangle appears as true length, and then rotate this view into an edge view. Although this appears to be complex, in reality, it is the simpler construction of Fig. A.10 and Fig. A.11 applied several times in sequence.

CONSTRUCTION (Fig. A.13)

1. Draw three projection lines joining the points *A*, *B*, and *C* in the plan and front views.
2. In order to show the true length of one side of the triangle, draw FL 2 parallel to that side. Any side will do; *AB* of the plan view is used here.
3. Projection lines perpendicular to FL 2 from the plan view and arcs from the front view established the corners of the triangle in Aux 1, where *AB* is

the true length of that side. The small circles mark the centers of the constructed arcs.

4. To rotate the triangle of Aux 1 into an edge view, draw FL 3 perpendicular to *AB*. Projection lines from Aux 1 and arcs from the plan view again locate the corners of the triangle in Aux 2; note that the line *AB* is in end view and therefore appears as a single point.
5. By rotating this view 90° the true shape of the triangle appears. With FL 4 parallel to the edge view, projection lines and arcs fix the corners of the triangle in Aux 3, the normal view.

THE METRICAL PROBLEMS

Distance from point to plane

PROBLEM

Given two views of a triangle *ABC* and a point *X* not in its plane. Construct the distance from the point to the plane.

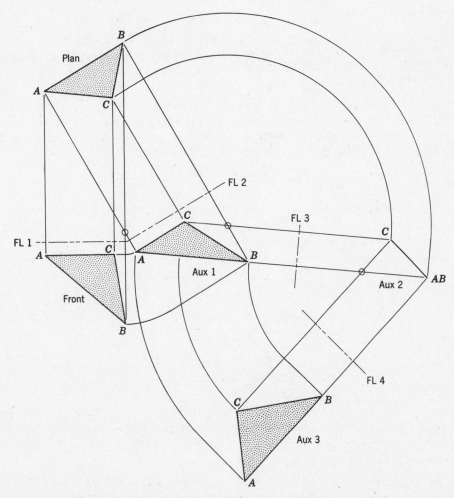

FIGURE A.13 Rotation of a plane figure into a normal view.

APPROACH

To determine this distance the figure must be rotated into an edge view, and the point rotated with it. This rotation of the plane is identical with the construction of the true shape of the plane figure (Fig. A.13).

CONSTRUCTION (Fig. A.14)

1. Draw projection lines to connect the points A, B, and C of the triangle and point X in space from the plan to the front views perpendicular to FL1.
2. Construct Aux 1 with FL 2 parallel to side AB of the plan view. Projection lines and arcs give a new view of the point, and the plane with AB as true length.
3. Construct Aux 2 with FL 3 perpendicular to AB of Aux 1. Lines and arcs from the plan view and Aux 1 give the required edge view.
4. The perpendicular distance, show dashed, from X to the plane of the edge view is the required distance from the point to the plane.

Angle between line and plane

PROBLEM

Given a plane represented by triangle ABC and a line DE inclined to this plane, find the angle between the plane and the line.

APPROACH

The angle between a plane and an intersecting line is measured in the plane containing the line which is perpendicular to the plane figure. This requires a series of rotations until this plane is in normal view.

CONSTRUCTION (Fig. A.15)

1. Projection lines are drawn to connect points A, B, and C which define the plane, and D and E of the line from the plan to the front view, perpendicular to FL 1.
2. Folding on FL 2 parallel to AB of the plan views, using projection lines and arcs, establishes Aux 1 of the line and the triangle with AB as true length.
3. Folding on FL 3 perpendicular to AB of Aux 1 gives the triangle in edge view; and the line.
4. Folding on FL 4 parallel to this edge view established Aux 4 showing the plane figure in its true shape and the line D as its own projection in this plane.
5. Folding on FL 5 parallel to DE established the normal view of the required plane; θ is the required angle.

Angle between two planes

PROBLEM

Given two intersecting planes defined by triangles ABC and ABD, determine the dihedral angle between them (Fig. A.16).

APPROACH

The dihedral angle between two planes is measured in the plane perpendicular to the line of

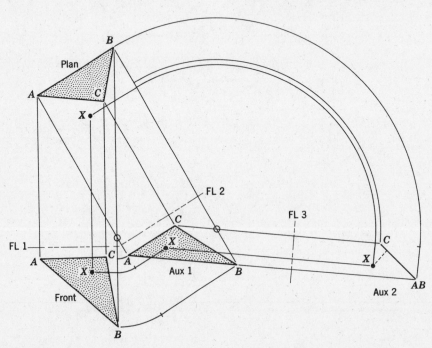

FIGURE A.14 Construction to determine the distance from a point to a plane.

intersection. To determine this angle an end view of this line must be obtained.

CONSTRUCTION (Fig. A.16)

1. Projection lines connecting the four points A, B, C and D, which define the two intersecting plane, are drawn perpendicular to FL 1.
2. Construct Aux 1 to show the line of intersection AB in true length.
3. With FL 2 perpendicular to AB, construct Aux 2 to show this line in end view.
4. The angle θ is the dihedral angle.

Distance between two lines

PROBLEM

Given two lines AB and CD, determine the shortest distance between them (Fig. A.17).

APPROACH

The shortest distance between two nonintersecting, nonparallel lines is measured along the perpendicular line common to both; this line has a fixed position in space. This relationship may be easily visualized by observing two pencils held in space.

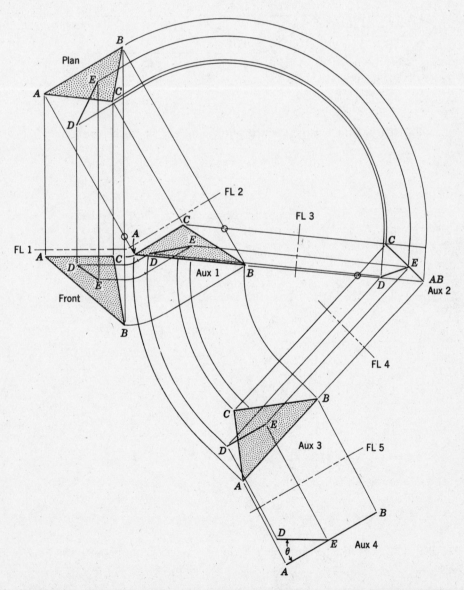

FIGURE A.15 Construction to determine the angle between a line and a plane.

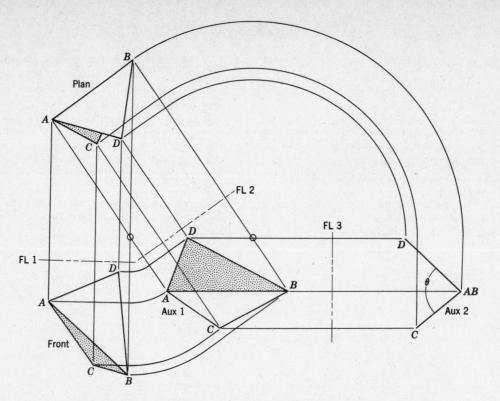

FIGURE A.16 Construction to determine the angle between two planes.

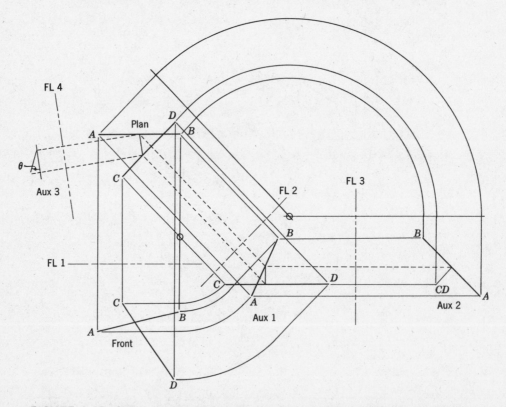

FIGURE A.17 Construction to determine the shortest distance between two lines in space.

CONSTRUCTION (FIG. A.17)

1. Projection lines are drawn connecting the ends of the line in the plan and front views.
2. Construct Aux 1 to show one of the lines in its true length; *CD* is used here.
3. Construct Aux 2 to show this line *CD* in end view. The perpendicular distance from the point *CD* to the line *AB* is the required distance.

4. If the location of this shortest distance line is needed, it is a simple matter to draw projection lines back through Aux 1 to the plan view (these lines are shown dashed). To determine the inclination of the line in this view, draw FL 4 parallel to its projection in the plan and with projection lines establish Aux 3. The hypotenuse of the triangle is the short line distance found in Aux 2, giving the angle of inclination θ.

APPENDIX

B

Trigonometric Functions

Degree	Sin	Tan		Degree	Sin	Tan		Degree	Sin	Tan	
0	.00000	.00000	90								
1	.01745	.01746	89	31	.51504	.60086	59	61	.87462	1.8040	29
2	.03490	.03492	88	32	.52992	.62487	58	62	.88295	1.8807	28
3	.05234	.05241	87	33	.54464	.64941	57	63	.89101	1.9626	27
4	.06976	.06993	86	34	.55919	.67451	56	64	.89879	2.0503	26
5	.08716	.08749	85	35	.57358	.70021	55	65	.90631	2.1445	25
6	.10453	.10510	84	36	.58779	.72654	54	66	.91355	2.2460	24
7	.12187	.12278	83	37	.60182	.75355	53	67	.92050	2.3559	23
8	.13917	.14054	82	38	.61566	.78129	52	68	.92718	2.4751	22
9	.15643	.15838	81	39	.62932	.80978	51	69	.93358	2.6051	21
10	.17365	.17633	80	40	.64279	.83910	50	70	.93969	2.7475	20
11	.19081	.19438	79	41	.65606	.86929	49	71	.94552	2.9042	19
12	.20791	.21256	78	42	.66913	.90040	48	72	.95106	3.0777	18
13	.22495	.23087	77	43	.68200	.93252	47	73	.95630	3.2709	17
14	.24192	.24933	76	44	.69466	.96569	46	74	.96126	3.4874	16
15	.25882	.26795	75	45	.70711	1.0000	45	75	.96593	3.7321	15
16	.27564	.28675	74	46	.71934	1.0355	44	76	.97030	4.0108	14
17	.29237	.30573	73	47	.73135	1.0724	43	77	.97437	4.3315	13
18	.30902	.32492	72	48	.74314	1.1106	42	78	.97815	4.7046	12
19	.32557	.34433	71	49	.75471	1.1504	41	79	.98163	5.1446	11
20	.34202	.36397	70	50	.76604	1.1918	40	80	.98481	5.6713	10
21	.35837	.38386	69	51	.77715	1.2349	39	81	.98769	6.3138	9
22	.37461	.40403	68	52	.78801	1.2799	38	82	.99027	7.1154	8
23	.39073	.42447	67	53	.79864	1.3270	37	83	.99255	8.1443	7
24	.40674	.44523	66	54	.80902	1.3764	36	84	.99452	9.5144	6
25	.42262	.46631	65	55	.81915	1.4281	35	85	.99619	11.430	5
26	.43837	.48773	64	56	.82904	1.4826	34	86	.99756	14.301	4
27	.45399	.50953	63	57	.83867	1.5399	33	87	.99863	19.081	3
28	.46947	.53171	62	58	.84805	1.6003	32	88	.99939	28.636	2
29	.48481	.55431	61	59	.85117	1.6643	31	89	.99985	57.290	1
30	.50000	.57735	60	60	.86603	1.7321	30	90	1.0000		0
	Cos	Cot	Degree		Cos	Cot	Degree		Cos	Cot	Degree

REFERENCES

Anderson, E.M., 1951, *The Dynamics of Faulting:* 2nd ed., Edinburgh, Oliver and Boyd, 206 p.

Argand, E., 1911, Les Nappes de recouvrement des Alpes Pennines et leurs prolongements structuraux: *Beiträge zur Geologischen Karte der Schweiz*, nue Folge 31, p. 1-26.

Badgley, P.C., 1959, *Structural Methods for the Exploration Geologist:* New York, Harper & Row, 280 p.

Balk, Robert, 1937, Structural behavior of igneous rocks: *Geol. Soc. America*, Mem. 5, 177 p.

Barnes, C.W., and R.W. Houston, 1969, Basement response to the Laramide orogeny at Coad Mountain, Wyoming: *Contrib. Geol.*, v. 8, p. 37-41.

Bayly, M.B., 1971, Similar folds, buckling and great-circle patterns: *Jour. Geology*, v. 79, p. 110-118.

Berthelsen, A., 1960, Structure contour maps applied in the analysis of double fold structures: *Geol. Rundschau*, v. 49, p. 459-466.

Berthelsen, A., E. Bondesen and S.B. Jensen, 1962, On the so-called Wildmigmatites: *Krystalinikum*, v. 1, p. 31-49.

Bucher, W.H., 1933, *The Deformation of the Earth's Crust:* Princeton, Princeton University Press, 518 p.

Busk, H.G., 1929, *Earth Flexures:* London, Cambridge University Press, 106 p.

Capps, S.R., 1919, The Kantishna Region, Alaska: *U.S. Geol. Surv. Bull.* 687, 116 p.

Carey, S.W., 1962, Folding: *J. Alberta Soc. Petrol. Geol.*, v. 10, p. 95-144.

Chaudhuri, A.K., 1972, Concise description of fold orientation: *Geol. Mag.*, v. 109, p. 231-233.

Compton, R.R., 1962, *Manual of Field Geology:* New York, Wiley, 378 p.

Compton, R.R., 1966, Analyses of Pliocene-Pleistocene deformation and stresses in northern Santa Lucia Range, California: *Geol. Soc. America Bull.*, v. 77, p. 1361-1380.

Connolly, H.J.C., 1936, A contour method of revealing some ore structures: *Econ. Geol.*, v. 31, p. 259-271.

Crowell, J.C., 1959, Problems of fault nomenclature: *Am. Assoc. Petroleum Geologists Bull.*, v. 43, p. 2653-2674.

Cruden, D.M., 1971, Traces of a lineation on random planes: *Geol. Soc. America Bull.*, v. 82, p. 2303-2305.

Cruden, D.M., and H.A.K. Charlesworth, 1972, Observations on the numerical determination of axes of cylindrical and conical folds: *Geol. Soc. America Bull.*, V. 83, p. 2019-2024.

Dahlstrom, C.D.A., 1969a, Balanced cross sections: *Canadian Jour. Earth. Sci.*, v. 6, p. 743-757.

Dahlstrom, C.D.A., 1969b, The upper detachment in concentric folding: *Bull. Canad. Petrol. Geol.*, v. 17, p. 326-346.

Dallmus, K.F., 1958, Mechanics of basin evolution and its relation to the habitat of oil in the basin: *in* Weeks, L.G., ed., Habitat of oil: *Amer. Assoc. Petroleum Geologists*, p. 883-931.

De Sitter, L.U., 1964, *Structural Geology:* 2nd ed., New York, McGraw-Hill, 551 p.

De Sitter, L.U., and H.J. Zwart, 1960, Tectonic development in supra- and infrastructes of a mountain chain: *21st International Geol. Congr.*, Pt. 18, p. 248-256.

Den Tex, E., 1954, Stereographic distinction of linear and planar structures from apparent lineations in random exposure planes: *J. Geol. Soc. Australia*, v. 1, p. 55-66.

Denness, Bruce, 1970, A method of contouring polar diagrams using curvilinear counting cells: *Geol. Mag.*, v. 107, p. 61-65.

Denness, Bruce, 1972, A revised method of contouring stereograms using curvilinear cells: *Geol. Mag.*, v. 109, p. 157-163.

Dennis, J.G. 1967, International Tectonic Dictionary: *Am. Assoc. Petroleum Geologist*, Mem. 7, 196 p.

Dickinson, W.R., 1966, Structural relationships of San Andreas fault system, Cholame Valley and Castle Mountain Range: *Geol. Soc. America Bull.*, v. 77, p. 707-726.

Doblin, J., 1956, *Perspective:* New York, Whitney, 68 p.

Donath, F.A., 1962, Analysis of Basin-Range structure, South-Central Oregon: *Geol. Soc. America Bull.*, v. 73, p. 1-16.

Donath, F.A., 1963, Fundamental problems in dynamic structural geology: *in* T.W. Donnelly, ed., *Earth Sciences* — problems and progress in current research: Chicago, Univ. Chicago Press, p. 83-103.

Donath, F.A., 1964, Strength variation and deformational behavior in anisotropic rocks: *in* Judd, W.R., ed., *State of stress in the earth's crust*, New York, American Elsevier, p. 281-297.

Dunnet, D., 1969, A technique of finite strain analysis using elliptical particles: *Tectonophysics*, v. 7, p. 117-136.

Earle, K.W., 1934, *Dip and Strike Problems Mathematically Surveyed:* London, Murby, 126 p.

Elliott, D., 1970, Determination of finite strain and initial shape from deformed elliptical objects: *Geol. Soc. America Bull.*, v. 81, p. 2221-2236.

Ferguson, H.G., and S.W. Muller, 1949, Structural geology of the Hawthorne and Tonopah Quadrangles, Nevada: *U.S. Geol. Surv.* Prof. Paper 216, 55 p.

Fisher, D.J., 1938, Problem of two tilts and the stereographic projection: *Am. Assoc. Petroleum Geologists Bull.*, v. 22, p. 1261-1271.

Fleuty, M.J., 1964, The description of folds: *Proc. Geol. Assoc.*, v. 75, p. 461-492.

Garland, G.D., 1971, *Introduction to geophysics - mantle, core and crust*: Philadelphia, Saunders, 420 p.

Gill, J.E., 1971, Continued confusion in the classification of faults: *Geol. Soc. America Bull.*, v. 82, p. 1389-1392.

Gilluly, J., A.C. Waters and A.O. Woodford, 1968, *Principles of Geology*: 3rd Ed., San Francisco, Freeman, 687 p.

Goguel, J., 1952, *Traité de tectonique:* Paris, Masson et Cie., 383 p.

Goguel, J., 1962, *Tectonics:* San Francisco, Freeman, 384 p.

Handin, John, 1969, On the Coulomb-Mohr failure criterion: *Jour. Geophys. Res.*, v. 74, p. 5343-5348.

Hansen, E., 1971, *Strain Facies:* New York, Springer-Verlag, 207 p.

Higgins, C.G., 1962, Reconstruction of flexure folds by concentric arc method: *Am. Assoc. Petroleum Geologists Bull.*, v. 46, p. 1737-1739.

Hill, M.L., 1963, Role of classification in geology: *in* C.C. Albritton, ed., *The fabric of geology*: Reading, Mass., Addison-Wesley, p. 164-174.

Holister, G.S., 1967, *Experimental stress analysis*: London, Cambridge University Press, 322 p.

Hopwood, T., 1968, Derivation of a coefficient of degree of preferred orientation from contoured fabric diagrams: *Geol. Soc. America Bull.*, v. 79, p. 1651-1654.

Hsü, K.J., 1968, Principles of mélanges and their bearing on the Franciscan-Knoxville paradox: *Geol. Soc. America Bull.*, v. 79, p. 1063-1074.

Hughes, R.J., 1960, A derivation of Earle's formula for the calculation of true dip: *Southeastern Geology*, v. 2, p. 43-48

Jaeger, J.C., 1969, *Elasticity, Fracture and Flow:* London, Methuen, 268 p.

Jaeger, J.C., and N.G.W. Cook, 1969, *Fundamentals of Rock Mechanics:* London, Methuen, 513 p.

Johnson, A.M., 1970, *Physical processes in geology*: San Francisco, Freeman, Cooper & Co., 577 p.

Kalsbeek, F., 1963, A hexagonal net for the counting-out and testing of fabric diagrams: *Neues Jahrbuch für Mineralogie*, Monatshefte, v. 7, p. 173-176.

King, P.B., 1969, Tectonic map of North America: *U.S.Geol. Survey*.

Kottlowski, F.E., 1965, *Measuring Stratigraphic Sections:* New York, Holt, Rinehart & Winston, 253 p.

Kranck, E.H., 1953, Interpretation of gneiss structures with special reference to Baffin Island: *Proc. Geol. Assoc. Canada*, v. 6, p. 59-68.

Kupfer, D.H., 1966, Accuracy in geologic maps: *Geotimes*, v. 10, no. 7, p. 11-14.

Leney, G.W., 1963, A new nomogram for the solution of problems in structural and economic geology: *Econ. Geol.*, v. 58, p. 1326-1339.

LeRoy, L.W., and J.W. Low, 1954, *Graphic problems in petroleum geology*: New York, Harpers & Brothers, 238 p.

Levorsen, I.A., 1960, *Paleogeologic Maps:* San Francisco, Freeman, 174 p.

Lobeck, A.K., 1958, *Block Diagrams:* Amherst, Emerson-Trussell, 212 p.

Low, J.W., 1957, *Geologic Field Methods:* New York, Harpers & Row, 489 p.

Lowe, K.E., 1946, A graphic solution for certain problems of linear structures: *Am. Min.*, v. 31, p. 425-434.

Lyons, M.S., 1964, Interpretation of planar structure in drill-hole cores: *Geol. Soc. America Spec. Paper* 78, 66 p.

Mackin, J.H., 1950, The down-structure method of viewing geologic maps: *Jour. Geology*, v. 58, p. 55-72.

Mackin, J.H., 1962, Structure of the Glenarm series in Chester County, Pennsylvania: *Geol. Soc. America Bull.*, v. 73, p. 403-410.

Mandelbaum, H., and J.T. Sanford, 1951, Table for computing thickness of strata measured in traverse or encountered in bore hole: *Geol. Soc. America Bull.*, v. 63, p. 765-776.

Marsh, O.T., 1960, A rapid and accurate contour interpolator: *Econ. Geol.*, v. 55, p. 1555-1560.

Matthews, P.E., R.A.B. Bond, and J.J. Van den Berg, 1971, Analysis and structural implications of a kinematic model of similar folding: *Tectonophysics*, v. 12, p. 129-154.

McIntyre, D.B., and L. E. Weiss, 1956, Construction of block diagrams to scale in orthographic projection: *Proc. Geol. Assoc.*, v. 67, p. 145-155.

McKinstry, H., 1961, Structure of the Glenarm Series in Chester County, Pennsylvania: *Geol. Soc. America Bull.*, v. 72, p. 557-578.

Mogi, Kiyoo, 1971, Fracture and flow of rocks under high triaxial compression: *Jour. Geophys. Res.*, v. 76, p. 1255-1265.

Nevin, C.M., 1949, *Principles of Structural Geology:* 4th ed., New York, Wiley 410 p.

O'Driscoll, E.S., 1962, Experimental patterns in superposed similar folding: *Jour. Alberta Soc. Petrol. Geol.*, v. 10, p. 145-167.

O'Driscoll, E.S., 1964, Cross fold deformation by simple shear: *Econ. Geol.*, v. 59, p. 1061-1093.

Oertel, G., 1962, Extrapolation in geologic fabrics: *Geol. Soc. America Bull.*, v. 73, p. 325-342.

Palmer, H.S., 1918, New graphic method for determining the depth and thickness of strata and the projection of dip: *U.S. Geol. Surv.* Prof. Paper 120-C, p. 123-129.

Pierce, W.G., 1966, Jura tectonics as a décollement: *Geol. Soc. America Bull.*, v. 77, p. 1265-1276.

Phillips, F.C., 1963, *Introduction to Crystallography:* New York, Wiley, 340 p.

Phillips, F.C., 1971, *The Use of Stereographic Projection in Structural Geology:* Third ed., London, Edward Arnold, 90 p.

Potter, P.E., and F.J. Pettijohn, 1963, *Paleoccurents and Basin Analysis:* New York, Academic Press, 296 p.

Price, N.J., 1966, *Fault and Joint Development in Brittle and Semi-Brittle Rock:* Oxford, Pergamon, 176 p.

Price, N.J., 1970, Laws of rock behavior in the Earth's crust: *in* Somerton, W.H., ed., Rock mechanics — theory and practice, Proc. 11th Symposium on Rock Mechanics, *Am. Inst. Min. Met. Petrol. Eng.*, New York, p. 3-23.

Ragan, D.M. 1969a, Introduction to concepts of two-dimensional strain and their application with the use of card-deck models: *Jour. Geol. Ed.*, v. 17, p. 135-141.

Ragan, D.M., 1969b, Structures at the base of an ice fall: *Jour. Geology*, v. 77, p. 647-667.

Ragan, D.M., and M.F. Sheridan, 1972, Compaction of the Bishop Tuff, California: *Geol. Soc. America Bull.*, v. 83, p. 95-106.

Ramsay, J.G., 1961, The effects of folding upon the orientation of sedimentation structures: *Jour. Geology*, v. 69, p. 84-100.

Ramsay, J.G., 1962, Interference patterns produced by the superposition of folds of similar type: *Jour. Geology*, v. 70, p. 466-481.

Ramsay, J.G., 1963, Structure and metamorphism of the Moine and Lewisian Rocks of the Northwest Caledonides, *in* M.R.W. Johnson, and F.H. Stewart, eds., *The British Caledonides:* Edinburgh, Oliver and Boyd, p. 143-175.

Ramsay, J.G., 1964, The uses and limitations of Beta-Diagrams and Pi-Diagrams in the geometrical analysis of folds: *Quart. J. Geol. Soc.*, v. 120, p. 435-454.

Ramsay, J.G., 1967, *Folding and Fracturing of Rocks:* New York, McGraw-Hill, 568 p.

Ramsay, J.G., 1969, The measurement of strain and displacement in orogenic belts: *in* Kent, P.E., G.E. Satterthwaite, and A.M. Spencer, eds., Time and place in orogeny, Spec. Publ. No. 3, *Geol. Soc. London*, p. 43-79.

Rickard, M.J., 1971, A classification diagram for fold orientations: *Geol. Mag.*, v. 108, p. 23-26.

Rickard, M.J., 1972, Fault classification-discussion: *Geol. Soc. America Bull.*, v. 83, p. 2545-2546.

Rutten, M.G., 1969, *The geology of western Europe*: Amsterdam, Elsevier Publishing Co., 520 p.

Satin L., 1960, Apparent-dip computer: *Geol. Soc. America Bull.*, v. 71, p. 231-234.

Schryver, K., 1966, On the measurement of the orientation of axial plane of minor folds: *Jour. Geology*, v. 74, p. 83-84.

Schweinfurth, S.P., 1969, Contour finder—inexpensive device for rapid, objective contouring: *U.S. Geol. Survey* Prof. Paper 650-B, p. B147-B148.

Screven, R.W., 1963, A simple rule of V's of outcrop patterns: *Jour. Geol. Ed.*, v. 11, p. 98-100.

Secrist, M.H., 1941, Computing stratigraphic thickness: *Amer. J. Sci.*, v. 239, p. 417-420.

Spencer, A.B., and P.S. Claubaugh, 1967, Computer program for fabric diagrams: *Am. J. Sci.*, v. 265, p. 166-172.

Stauffer, M.R., 1966, An empirical-statistical study of three dimensional fabric diagrams as used in structural analysis: *Canadian Jour. Earth Sci.*, v. 3, p. 473-498.

Stauffer, M.R., 1968, The tracing of hinge-line ore bodies in areas of repeated folding: *Canadian Jour. Earth Sci.*, v. 5, p. 69-79.

Suter, H.H. 1947, Exaggeration of vertical scale of geologic sections: *Am. Assoc. Petroleum Geologists Bull.*, v. 31, p. 318-339.

Ten Haaf, E., 1967, Apparent-dip protractor: *Geol. Rundschau*, v. 56, p. 83-84.

Timoshenko, S., and J.N. Goodier, 1951, *Theory of Elasticity:* New York, McGraw-Hill, 506 p.

Turner, F.J., and L.E. Weiss, 1963, *Structural Analysis of Metamorphic Tectonites*: New York, McGraw-Hill, 545 p.

Verhoogen, John, F.J. Turner, L.E. Weiss, Clyde Wahrhaftig, and W.S. Fyfe, 1970, *The earth*: New York, Holt, Rinehart and Winston, 748 p.

Warner, F.M., and M. McNeary, 1959, *Applied Descriptive Geometry*: 5th ed., New York, McGraw-Hill, 243 p.

Warner, Jeffery, 1969, FORTRAN IV program for construction of pi diagrams with Univac 1108 computer: *Computer Contrib.* 33, Kansas State Geol. Survey, 38 p.

Watson, G.S., 1970, Orientation statistics in the earth sciences: *Bull. Geol. Inst.*, Univ. Uppsala, New Ser., v. 2, p. 73-89.

Wegmann, C.E., 1929, Beispiele tektonischer Analysen des Grundgebirges in Finnland: *Bulletin de la Commission géologique de Finlande*, no. 87, p. 98-127.

Wellman, H.G., 1962, A graphic method for analysing fossil distortion caused by tectonic deformation: *Geol. Mag.*, v. 99, p. 348-352.

Wentworth, C.K., 1930, The plotting and measurement of exaggerated cross-sections: *Econ. Geol.*, v. 25, p. 827-831.

White, W.S., 1946, A contangent ruler for simplifying graphic solutions of problems in structural geology: *Econ. Geol.*, v. 41, p. 539-545.

Wilson, G., 1961, Tectonic significance of small scale structures, and their importance to the geologist in the field: *Annales de la Société Géologique de Belgique*, v. 84, p. 423-548.

Wilson, G., 1967, The geometry of cylindrical and conical folds: *Proc. Geol. Assoc.*, v. 78, p. 179-210.

Wright, F.E., 1911, *The methods of petrographic-microscope research*: Carnegie Institution of Washington, Publ. No. 158.

Zimmer, P.W., 1963, Orientation of small diameter drill core: *Econ. Geol.*, v. 58, p. 1313-1325.

BIBLIOGRAPHY OF GEOMETRICAL TECHNIQUES

Badgley, P.C., 1959, *Structural Methods for the Exploration Geologist:* Harper and Brothers, New York, 280 p.

Bennison, G.M., 1965, *An Introduction to Geological Structures and Maps:* Edward Arnold, London, 64 p.

Billings, M.P., 1972, *Structural Geology:* Prentice-Hall, New Jersey, 514 p. (p. 494-588)

Blyth, F.G., 1965, *Geological Maps and Their Interpretation:* Edward Arnold, London, 48 p.

Dennis, J.G., 1972, *Structural Geology:* Ronald Press, New York 532 p. (p. 49-68)

Dennison, J.M., 1968, *Analysis of Geologic Structures:* W.W. Norton, New York, 209 p.

Donn, W.L., and J.A. Shimer, 1958, *Graphic Methods in Structural Geology:* Appleton-Century-Crofts, New York, 180 p.

Gwinner, M.P., 1965, *Geometrische Grundlagen der Geologie:* E. Schweizerbart'sche Verlagsbuchhandlung, Stuttgart, 154 p.

Higgs, D.V., and George Tunnell, 1959, *Angular Relations of Lines and Planes:* Wm. C. Brown, Dubuque, Iowa, 43 p.

LeRoy, L.W., and J.W. Low, 1953, *Graphic Problems in Petroleum Geology:* Harper and Brothers, New York, 238 p.

Nevin, C.M., 1949, *Principles of Structural Geology:* Fourth ed., John Wiley, New York, 410 p. (p. 329-399)

Phillips, F.C., 1971, *The Use of Stereographic Projection in Structural Geology:* Third edition, Edward Arnold, London, 90 p.

Platt, J.I., and John Challinor, 1968, *Simple Geological Structures:* Fourth ed., Thomas Murby, London, 56 p.

Russell, W.L., 1955, *Structural Geology for Petroleum Geologists:* McGraw-Hill, New York, 427 p. (geometrical material widely scattered).

Simpson, Brian, 1968, *Geological Maps:* Pergamon Press, Oxford, 98 p.

Spencer, E.W., 1969, *Introduction to the Structure of the Earth:* McGraw-Hill, New York, 597 p. (p. 555-578)

Thomas, J.A.G., 1966, *An Introduction to Geologic Maps:* Thomas Murby, London, 64 p.

Verhoogen, John, F.J. Turner, L.E. Weiss, Clyde Wahrhaftig, and W.S. Fyfe, *The Earth—an Introduction to Physical Geology:* Holt, Rinehart and Winston, New York, 748 p. (p. 709-728)

Index

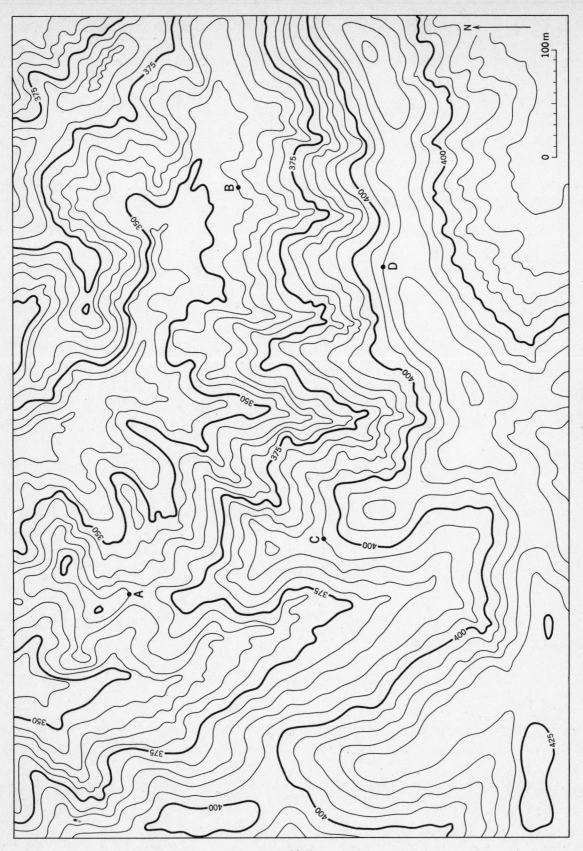

X-1

FIGURE X3.3

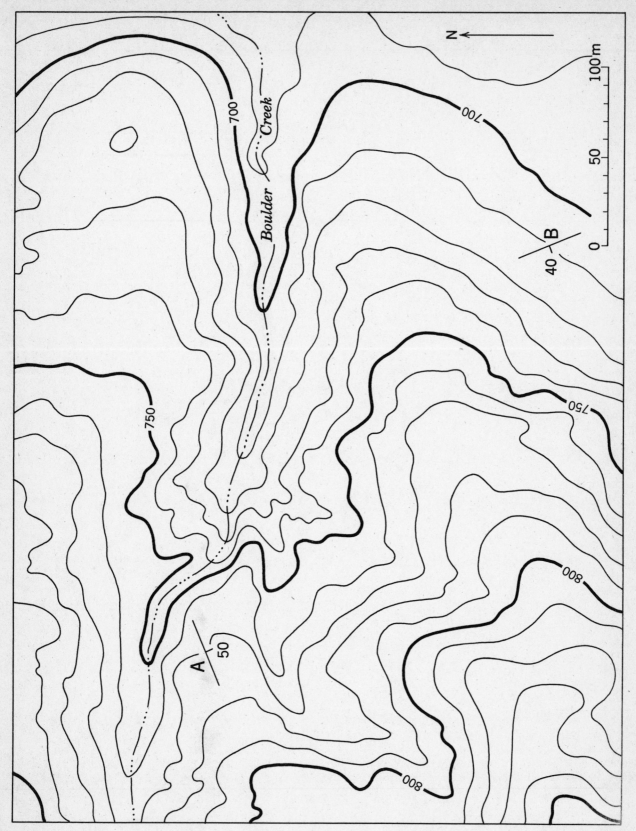

FIGURE X 4.1

FIGURE X5.1

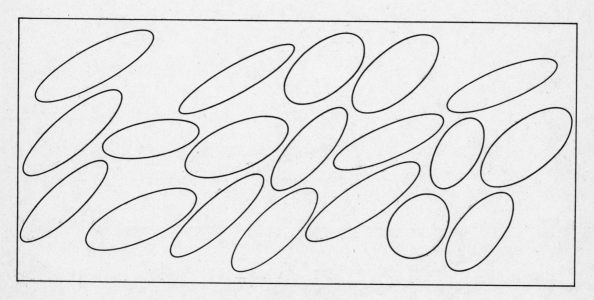

FIGURE X5.2

X–3

FIGURE X7.1

FIGURE X9.1

FIGURE X10.1

X–6

Scale 1 : 2000

FIGURE X10.2

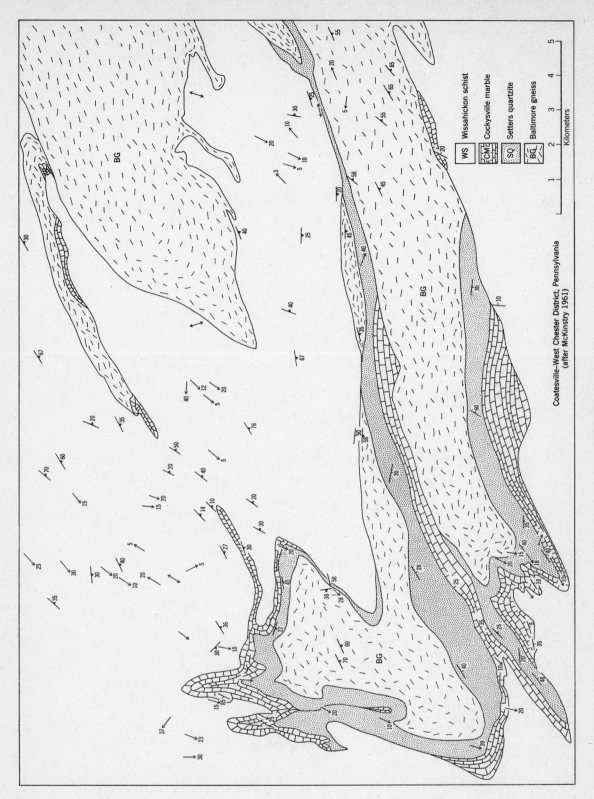

Coatesville–West Chester District, Pennsylvania
(after McKinstry 1961)

FIGURE X10.3

WS Wissahickon schist

CM Cockysville marble

SQ Setters quartzite

BG Baltimore gneiss

Kilometers
1 2 3 4 5

FIGURE X18.1

X-9

FIGURE X18.2

X-10

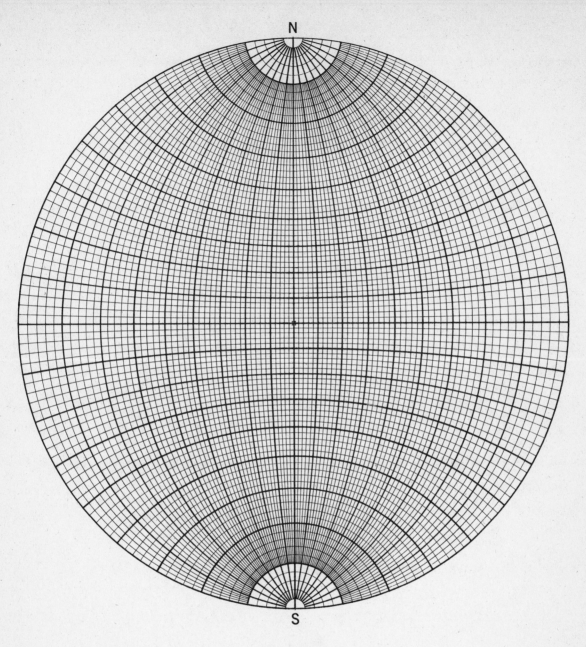

N

S

WULFF NET

X-11

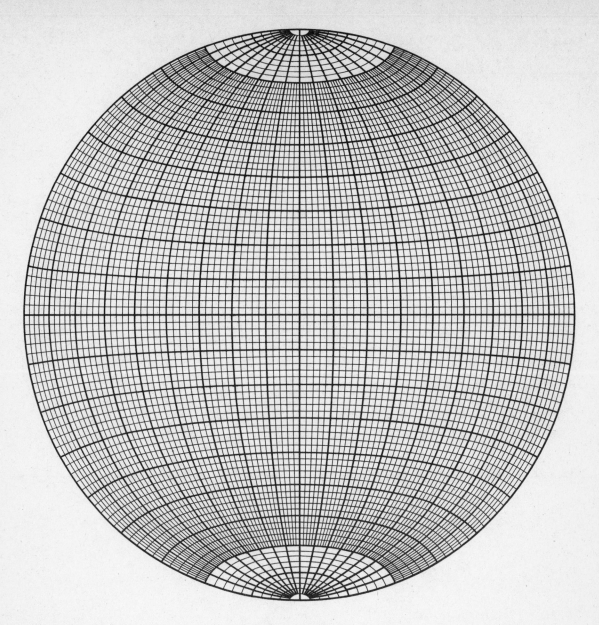

SCHMIDT NET

X-12

KALSBEEK COUNTING NET

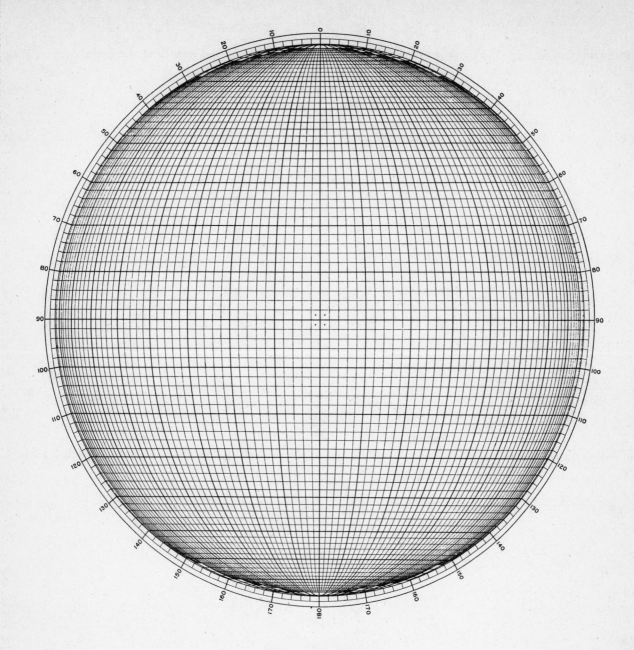

ORTHOGRAPHIC NET

X-14